Marshall Fishwick, PhD

Probing Po[pular Culture]
On and Of[f the Internet]

Pre-publication
REVIEWS,
COMMENTARIES,
EVALUATIONS . . .

"**M**arshall Fishwick's *Probing Popular Culture* is a learned and, at
the same time, readable volume. If you
had any illusions about popular culture
icons such as Paul Bunyan, Henry Ford,
or Joe Magarac, they'll be gone after
reading this book.

Fishwick strengthens his book with
a section on 'other voices,' including a
chapter by Ray B. Browne, one of the
founders and stalwarts of popular culture studies. In that same section, a high
school teacher adds comments about
the approach of teenagers and teachers
to new technology, and an anthropologist contributes on interesting piece on
'the realm of Splogia.'

As in his other work, Fishwick
roams widely over art, architecture, religion, politics, music, and technology.
For those who have an interest in using
this material in class or in going further
in the subjects covered, Fishwick includes an appendix on electronic resources, a good section on further reading, and notes for each of the chapters
covered."

Marvin Wachman, PhD
President Emeritus, Temple University

Michigan State University

More pre-publication
REVIEWS, COMMENTARIES, EVALUATIONS . . .

"In the early years of a new century, it is appropriate to try to make sense of our dynamic popular culture, and no one is better equipped to do this than Marshall Fishwick. One of the founders of the popular culture movement, Fishwick has been at the center of cultural study for more than five decades, providing insightful commentary on the things that matter most in our lives: our history, heroes, lifestyles, religion, arts, and technologies. In *Probing Popular Culture*, Fishwick demonstrates his grasp of how the various elements of our popular culture co-exist, sometimes smoothly and sometimes at odds, how they have developed over time, and what they mean. Impressive in its sweep, the book leaves no aspect of our culture unexamined.

Fishwick looks both admiringly and critically at the culture of the people, reminding us that 'popular culture is at the heart of revolutions.' He urges all of us not to dismiss its importance, and after reading this book, we never will. Written in a lively, readable style characterized by rhetorical questions, *Probing Popular Culture* engages readers from the start, compelling them to think."

Kathy Merlock Jackson, PhD
Editor, *The Journal of American Culture;*
Past-President, American Culture
Association

"Marshall Fishwick dissects vernacular folkways throughout the Global Village—on and off the Internet, in and out of bounds, up and down the pike. In the shelfful of books he has published over the past five de-

cades, Fishwick helped to make popular the study of popular culture. In his new book he probes the immediate past —the twentieth century—asking a thousand questions of it that the twenty-first century will need to answer."

Charles Boewe, PhD
Former Director of Fulbright Foundations
in Iran, Pakistan, and India

"This book offers an extraordinary range of information, ideas, and insights. Fishwick makes it clear that this is a book about popular culture, and early in the book he announces his intent: 'My goal is simple: to set minds working and tongues wagging.' Surely in this he succeeds."

Harry B. Adams, MDiv, HonDD
Horace Bushnell Professor Emeritus
of Christian Nurture,
Yale University Divinity School

"Marshall Fishwick has done it again. *Probing Popular Culture* is an extraordinary feat. An exhilarating collection of short, vignette-like chapters with an incredible range, he has established that there is a new form of culture, one that combines great technical sophistication with a Rude Boy atmosphere. Fishwick's grasp of the subject is at once awe-inspiring and magnetic. Get this book."

Daniel Walden, PhD
Professor Emeritus of American Studies,
English and Comparative Literature,
Penn State University

Probing Popular Culture
On and Off the Internet

THE HAWORTH PRESS
Titles of Related Interest

Great Awakenings: Popular Religion and Popular Culture by Marshall W. Fishwick

Popular Culture: Cavespace to Cyberspace by Marshall W. Fishwick

Rock Music in American Popular Culture: Rock 'n' Roll Resources by B. Lee Cooper and Wayne S. Haney

Rock Music in American Popular Culture II: More Rock 'n' Roll Resources by B. Lee Cooper and Wayne S. Haney

Rock Music in American Popular Culture III: More Rock 'n' Roll Resources by B. Lee Cooper and Wayne S. Haney

Barbershops, Bullets, and Ballads: An Annotated Anthology of Underappreciated American Music Jewels, 1865-1918 edited by William E. Studwell and Bruce R. Schueneman

Circus Songs: An Annotated Anthology by William E. Studwell, Charles P. Conrad, and Bruce R. Schueneman

The Classic Rock and Roll Reader: Rock Music From Its Beginnings to the Mid-1970s by William E. Studwell and David F. Lonergan

The Big Band Reader: Songs Favored by Swing Era Orchestras and Other Popular Ensembles by William E. Studwell and Mark Baldin

Popular Culture in a New Age by Marshall W. Fishwick

The Magic Window: American Television, 1939-1953 by James A. Von Schilling

Dictionary of Toys and Games in American Popular Culture by Frederick J. Augustyn Jr.

Basketball in America: From the Playgrounds to Jordan's Game and Beyond edited by Bob Batchelor

The Spectacle of Isolation in Horror Films: Dark Parades by Carl Royer and Diana Royer

Probing Popular Culture
On and Off the Internet

Marshall Fishwick, PhD

The Haworth Press®
New York • London • Oxford

The Haworth Press, Inc., 10 Alice Street, Binghamton, NY 13904-1580.

Portions of the material in this book have appeared, in somewhat different form, in the journals *Yale Review, Saturday Review,* and *Western Folklore,* and from Public Affairs Press, David McKay Company, The Haworth Press, and Thompson Learning Custom Publishing, and are published with permission.

Popular Culture Wheel. Browne, Ray B., and Marshall W. Fishwick, eds. *Symbiosis: Popular Culture and Other Fields.* © 1988. Reprinted by permission of The University of Wisconsin Press.

Cover design by Lora Wiggins.

Library of Congress Cataloging-in-Publication Data

Fishwick, Marshall William.
    Probing popular culture : on and off the Internet / Marshall Fishwick.
      p. cm.
    Includes bibliographical references and index.
    ISBN 0-7890-2132-3 (case : alk. paper) — ISBN 0-7890-2133-1 (soft : alk. paper)
    1. Popular culture—United States. 2. United States—Civilization—1970- 3. Technology and civilization. 4. Civilization, Modern—1950-. 5. Twenty-first century. 6. Postmodernism. 7. Civilization, Modern—21st century. 8. Popular culture—Forecasting. I. Title.
E169.12.F643 2004
306'.0973—dc22
                                                                                    2003017999

For all my students,
who over several student
generations, have taught me
more than I have taught them,
and who keep in touch.

I grope, I probe, I listen, I test until the tumblers fall and I'm in.

Marshall McLuhan

PROBE: from Latin, *probare*—to try or test. In late Latin, also an examination. To search into, to explore, to discover or ascertain something; to pierce or to penetrate.

The Oxford English Dictionary

Your question has probed through to the pith of our belief.

George Eliot, 1878

CONTENTS

ABOUT THE AUTHOR

Marshall Fishwick, PhD, Professor of Interdisciplinary Studies and Director of the American Studies and Popular Culture programs at Virginia Tech in Blacksburg, Virginia, holds several honorary degrees and teaching awards. Dr. Fishwick is the author of *Heroes and Superheroes*; *Symbolism*; *Revitalizing the Humanities*; *An American Mosaic: Rethinking American Culture History, Great Awakenings: Popular Religion and Popular Culture*; *Popular Culture: Cavespace to Cyberspace*; and *Popular Culture in a New Age* as well as many others. Fishwick serves as advisory editor to the *Journal of Popular Culture* and the *Journal of American Culture* and is cofounder of the journal *International Popular Culture*. In 2000, Dr. Fishwick authored a "Millennium Edition" of *Go and Catch a Falling Star* as well as articles about the millennium and the controversial "global village."

In 1998, he was honored by the American Culture Association (ACA) with a Lifetime Achievement Award. Dr. Fishwick is cofounder of the Popular Culture Association (PCA) and has served as its president. A Fulbright Distinguished Professor in Denmark, Italy, Germany, Korea, and India, he helped establish the American Studies Research Center in Hyderabad, India, which now houses the largest collection of American books in Asia. Dr. Fishwick has recently been appointed Senior Editor of Haworth's Popular Culture & American Studies book program.

CONTRIBUTORS

Ray B. Browne helped found the Popular Culture Association and also founded the *Journal of Popular Culture*. He has written and edited dozens of books and articles and is a major voice in academia.

James Combs retired from teaching early to become a full-time author. He has written many books and articles and continues to teach a seminar on popular films. He lives near Abingdon, Virginia.

Katherine Lynde is an enthusiastic high school teacher, and one with major interest in popular culture and pedagogy. She teaches honor sections in Blacksburg High School in Virginia.

Foreword

Coleridge's Ancient Mariner had a "glittering eye," a term that fascinated me for years, since I had never seen such a thing and couldn't imagine what it looked like. Then one day I met Marshall Fishwick. I was in my junior year at Washington and Lee when Marshall arrived as a young teacher fresh out of the American Studies doctoral program at Yale. One look at him in the classroom—and no question about it, there it was, he had it, the Mariner's orb. It blazed away. It dazzled. It lit up courses such as no one had ever encountered at Washington and Lee or any other school.

Coleridge's Mariner was a monomaniac, determined to tell everybody a single story. Marshall was the reverse, an omnimaniac. His scholarly interest knew no bounds, and he was determined to *teach it all.* His pièce de résistance was an extraordinary omnibus course in American intellectual history. He covered American philosophical thought, American psychological theory, theology, historiography, scientific method, the works—all with the glittering eye and a glistening smile that seemed to come not so much from down-home happiness as from the exhilaration of the quest for knowledge and the challenge of getting it through our skulls.

He was the most magnetic teacher I had ever known, and as soon as I left Washington and Lee I headed for the American Studies program at Yale in hopes of acquiring the same Dionysian sweep myself. But of course Marshall was, to borrow Huey Long's phrase, sui generis. I have followed his career over the decades since then with a growing admiration, through his thirty books, and his innumerable articles, monographs, and newspaper columns. I have also made a point of keeping up with his amazing classroom performances. Lately I have gone to see him in action at Virginia Tech—and I have watched the glittering eye and the glistening smile entrancing another fortunate generation of undergraduates.

All the while, Marshall's interests have continued to expand, much the way old-fashioned (last year's) physicists believe the universe does. One of his latter-day pursuits has been the study of popular culture, a subject big enough and broad enough even for him. It was Matthew Arnold who coined the term *culture* as a word referring to the arts and literature. Culture, he believed (along with Max Weber and Hugo von Hoffmansthal), would play an increasingly important role in modern society. In fact, he argued, it had already given rise to a fourth social class, joining the traditional upper class ("the Barbarians," Arnold called them), the middle class ("the Philistines"—another Arnold coinage), and the lower class ("the Populace"—not even in the nineteenth century did intellectuals dare be so incorrect as to dream up funny names for the bottom dogs). To this new class, this fourth class, the "culture" class, he gave the cumbersome name of "the people of sweetness and light." ("Sweetness and light": yet another Arnold coinage, along with "commuter," meaning a poor drudge who goes back and forth from a small town to the big city in order to toil.) Sweetness and light were all that were required to join the new class, but to Arnold it was axiomatic that such dulcet and luculent creatures could be produced only by Oxford and Cambridge or their equivalents, if any. In other words, culture, while not the property of the merely moneyed classes (the Barbarians and the Philistines) nevertheless belonged to an elite.

In the United States, however, particularly in the second half century, people definitely not of the Oxford cut have developed a new form of "culture," one that often combines great technical sophistication with a Rude Boy mental atmosphere. Its works and its influence have become well nigh unavoidable. It is known as Popular Culture. It has become such a big part of modern life, one can no longer rule it outside the boundaries of scholarly attention. Bringing it inside has proved to be a severe test of academic rigor. It also demands scholars of unprecedented range, powers of synthesis, and, not least of all, energy. What is to be done?

Fellow students, I give you Marshall Fishwick.

Tom Wolfe

Preface

The people are a giant Atlas, carrying the world on their shoulders. How and why do they do it? To ask such questions will always be popular culture's first task.

Marshall Fishwick

Marshall Fishwick, a founding father of Popular Culture studies, has set the tone and pace of that major movement. This is a splendid summary of what he thinks and knows in our new millennium.

During the turmoil of the 1960s, he was an early pioneer in American Studies, receiving his doctorate at Yale University under Professor Ralph Gabriel. Fishwick began American Studies programs at Washington and Lee University and Lincoln University before joining with Professors Ray B. Browne and Russel B. Nye in 1967 to start the Popular Culture Association, which would both broaden and deepen cultural studies and break down the constricting boundaries of well-entranced departments. The idea caught fire and has burned brightly ever since. Two major journals—*The Journal of Popular Culture* and *The Journal of American Culture*—are the flagships, and scores of books from the Popular Press are the legacy. This book is a distillation of that pioneering and these triumphs.

A graduate of Mr. Jefferson's University of Virginia, Fishwick believes, with Jefferson, that if you give the people light, they will find their way. He also agrees that it is the great multitude for whom all really great things are done, said, suffered. The multitude desires the best of everything and in the long run is the best judge of it. Fishwick also knows there is truth in P. T. Barnum's famous cynical remark: "There's a sucker born every minute." He gives suckers a local habitation and a name.

In an earlier book, he explained, "Why not put our trust in Electronic Darwinism?" Rather than curtail or restrict information, let it all flow and then trust our great parents, Mother Nature and Father Time, to sort it all out. The silly and stupid must be discarded, the significant will be retained. In a democracy, we must trust the people to judge, as eventually they will. Here, indeed, is an ex-elitist's trust in the people.

Every thinking American is concerned about how the media are shaping our daily lives. Too often, the answers of scholars are delivered in jargon-ridden studies which—when closely examined—signify little or nothing. In contrast, *Probing Popular Culture* is an entertaining and mature work which will make sense to the college student, the general reader, and the specialist. In the Declaration of Independence, Thomas Jefferson talked about the "pursuit of happiness." At the time—and later—there have been varying interpretations of what he meant in that vague phrase. In this direct and informative overview, a scholarly founding father shows how popular culture studies have become a way to understand how ordinary people find meaning and joy in their lives. The result is nothing less than a better understanding of who we are and who we should become. I can think of no more humanistic endeavor—can you?

Peter Rollins

Acknowledgments

I owe so much to so many that I could fill a chapter—thank you to my teachers and benefactors. My family put my education above all else, and gave us opportunities my parents never had. I took the "classical course route" at a fine high school, and had excellent mentors at the University of Virginia, the University of Wisconsin, and Yale University, and I shall never forget them. Mine is a helpful profession.

Those that shaped me and set the tone of my life include Sally Lovelace, James Southall Wilson, Peters Rushton, A. K. Davis, Merle Curti, Ralph Gabriel, Arnold Toynbee, Whitney Griswald, and James G. Leyburn.

Marvin Wachman and Ray Browne, who still befriend and guide me, ventured into new fields, such as popular culture, and are dear in my heart.

Former students, such as Tom Wolfe, Roger Mudd, Carl Barnes, and Roy Matthews have shown me how a new generation makes new advances; thus do one's students become his teachers.

A number of critics and editors have helped me get the spoken word into print. Norman Cousins opened the pages of the *Saturday Review* to me, as have a series of generous editors at home and abroad.

At the Haworth Press, which has published my more recent books, I owe much to Patricia Brown, Peg Marr, and Bill Palmer. For my shortcomings I must take all the blame.

The author and the Colonel welcome you.

Introduction

Welcome to the Twenty-First Century

Fasten your seat belt. It's going to be a bumpy trip. It already has been, since the big ball came down in Times Square at midnight, January 1, 2000. No one expected then what lay ahead.

Has any age come in with the hip-hype hurrah of the New Millennium? On January 1, 2000, the whole world entered the celebration mode. Millions of dollars supported parades, light shows, special editions, concerts, extravaganzas, and fireworks lit up the skies from Fiji to Fredonia. *Bring on Utopia!*

After all, no one alive had ever welcomed a new millennium before; and no one would live to see the next one in 3000 A.D. The chance of a lifetime! Ring out wild bells! The opening of a New Age!

Nowhere was the celebration more long and lavish than in the United States. "We pulled out all the stops"; politicians, pollsters, the media, and people everywhere lit the skies and their homes on January 1, 2000. The dawn of a new millennium! After the euphoria, Americans returned to the task they relish most: electing a president and many other candidates at all levels. The ensuing presidential race was fierce, and (like eight other presidential elections before it) saw the candidate with the most popular votes (Al Gore) denied that honor by the electoral college. The Founding Fathers had made sure this was a Republic, rather than a Democracy. The U.S. Supreme Court, in a 7-2 decision on December 12, reversed a Florida Supreme Court decision ordering the hand recounts of thousands of votes, which gave George Bush his razor-thin election.

The year that had begun with free-flowing confetti ended with uncountable disputed ballots and pregnant chads.

The presidential election was so close that the final outcome depended on who had won Florida. The electoral apparatus apparently broke down, and the results were inconclusive. The fact that George Bush's brother was governor of Florida added a disquieting detail.

1

Was the election fair? Might the nightmare be repeated in the 2004 election?

For many, the contested election results have never been accepted. In many other areas the complacency of the 1990s evaporated, as much of the optimism and joy that heralded the New Day seemed to disappear. The Middle East was in shambles, with blood all over the ground where the Prince of Peace once walked. The West had failed to bring any workable truce in the equally bloody Balkans, and years of threatening Saddam Hussein and his rogue nation of Iraq proved futile when he drove the arms inspectors out of his country. All over the globe, weary disappointed peacekeepers were anxious to go home. Muslim anger mounted.

AIDS, Ebola, corruption, and tribal outbursts swept through Africa and Asia. In 2001 our economic bubble finally burst. Stock markets around the globe tumbled, as Bears chased the Bulls out of Wall Street and high-tech stocks reached all-time lows.

Still, most Americans seemed pleased after a decade of prosperity, and expected it to return quickly. Didn't we have the strongest economy in the world, and weren't these little setbacks part of the whole system? Was the bubble of the 1990s about to burst? Yes. It came like a thief, in what might turn out to be the determining event of our generation, even our century.

It burst with a terrible bang on September 11, 2001. On that day came the shock heard round the world: the attack on the World Trade Center and the Pentagon, by terrorist hijackers, with our own commercial planes, fueled for long flights and carrying thousands of gallons of high octane fuel. The result was thousands of deaths and injuries and the end of the myth of our invulnerability. Our two oceans, which had protected the homeland for centuries, could not protect us from this kind of attack that was launched in the skies above our own soil. Many likened 9/11 to the sneak attack on Pearl Harbor on December 7, 1941, which President Franklin D. Roosevelt had called "a day that would live in infamy." We shall revisit 9/11 later in this book.

If that date created thousands of victims, it also gave us thousands of new heroes, such as the firefighters, police officers, and thousands of volunteers who risked their lives attempting to save others. Stature of some politicians soared, most notably Mayor Giuliani of New York City and President George W. Bush.

Some have called post-9/11 America the land of the New Fear. Will we be attacked again? When? How? A poll in 2002 indicated that three out of four New Yorkers feared there would be another attack.

The events of 9/11 set off a ripple effect of problems and setbacks that would continue for many months, such as decline in consumer confidence, rising unemployment, and bankruptcies. By 2003 over 200,000 jobs were lost. Our continuing economic decline would bring the stock market, by October 2002, to its lowest point since the Great Depression.

On October 2, 2002, a window was shot out in Aspen Hill, Maryland, and a man was killed in nearby Wheaton. On a bloody October 3, five people, three men and two women, were ambushed and shot, four in Maryland, one in Washington, DC. A serial killer was on the loose. Ballistic tests indicated .223 caliber bullets from the same high-powered rifle were used. The rampage continued.

Panic gripped the area. Killings continued, reaching down into Virginia. The tarot card "Death" was found, inscribed "I am God." A later note demanded $10 million, then a phone call: "Your children are not safe." Panic spread, schools closed, stores emptied.

America's most extraordinary manhunt, involving thousands of police, troops, airplanes, and journalists ended on October 24. An army veteran John Allen Muhammad and a Jamaican teenager, Lee Boyd Malvo, thought to be Muhammad's stepson, were arrested and in 2003 were tried and convicted. In 2004 their final fate was still to be determined, pending the outcome of court appeals.

Not all our new century problems came from abroad, nor from terrorists. Some grew out of the very thing we cherished most, the world's most advanced and efficient technology. We led the Electronic Revolution—until the lights went out.

On a sultry summer afternoon—August 14, 2003—America was stunned by the biggest blackout in U.S. history. It rolled instantly across a swath of the northern United States and southern Canada, driving millions outdoors into rush-hour streets; shutting down nuclear power plants; trapping people in buildings and subways; and closing airports.

How could and did it happen? No one seemed to know. All President Bush could say was "We'll find out why it happened and we'll deal with the problem."

How long did the disaster take? It happened in about nine seconds, blocking 10 percent of electricity east of the Rocky Mountains. Could it be restored that fast? No, for if an outage is restored too quickly, it causes another system failure.

One recalls Henry Adams's great study: "The Dynamo and the Virgin." Once we put our faith in the church; now, Adams noted, we put it in electricity. The wire in our hand might not enlighten, but electrocute us. Technology might well prove to be the god that failed.

In December 2003 a rash of sniper killings erupted on Interstate 70 outside Columbus, Ohio. Again, an army of investigators had not found the killers as the year ended. Was this a portent of a new kind of problem for twenty-first-century America, and perhaps of the world?

Far away, on the idyllic island paradise of Bali, a peaceful tourist haven, on October 12, a pair of bombings turned paradise into an inferno, killing 202 people and wounding hundreds more. This worst attack in Indonesia's history turned suspicion to al-Qaida and an affiliated group, Jemaah Islamiyah. What frightened Americans was that the attacks were near the second anniversary of the al-Qaida-linked attack on the USS *Cole* off Yemen, which occurred October 13, 2000, and led to the closure of U.S. embassies and renewed terrorist alerts for Americans. The Global Village was exploding.

War clouds darkened the horizon. Congress approved military action against Iraq, despite much opposition. We stood on the brink of World War III.

Life in the twenty-first century finds us on a roller coaster: ups and downs to take our breath away and leave the whole world gasping. All the old truisms may be false; our two oceans offer no protection, and there is no place to hide. Everything nailed down is coming loose.

We live by instant issues, threats, posters, and protests. We seem to be for and against so many things that we find most issues muddied. Just what are the issues? Whose side are we really on? Why are our former allies defecting? Why are we so hated by so many?

Plagued by earthbound problems, we move into outer space. The costs are enormous, and so far, the results are questionable. Our astronauts transmit back pictures of earth—a small orb floating in a cosmic sea.

Why do so few people vote, or cope? Are we sacrificing civil rights for the terrorist threat? And are the terrorists succeeding not only here but around the world?

Where can we find new wellsprings of energy and faith? Can we find new heroes now that so many old ones are discredited? Can we clear out the junk not only from our attics and highways, but also from our minds? Civilize technopolis and make cities livable?

I do not imply that our end is near; but our utopian dream of turning the globe into a free-trade shopping mall, commanded by dollar diplomacy, monitored by computers and technology, may be yet another utopian dream. Evidence keeps mounting that nationalism, tribalism, and localism are growing stronger, not weaker. What happens, in the new electronic takeover, to the needs of primary groups, subgroups, and traditional communities? Look at the situation in Northern Ireland, the Balkans, Africa, the Middle East, and despair. Not only financial panic but balkanization are epidemic.

In short, our post–Cold War predictions and actions were too simplistic and superficial. They underestimated the power of tradition, local autonomy, and local memory. Most people in the world still live as did their ancestors, in tribal, local, or national patterns, and want to continue to live that way.

True, an elite has grabbed the technology (and often the profits) of the Information Age. But most people—the so-called "common folk"—expect to live and be buried close to where they were born. They cannot watch CBS or CNN to keep up with their local news.

But doesn't the spread and instant availability of world news unite people? Professor Colin Cherry at the University of London thinks not. It is grossly naive to assume that expanding and consolidating world communication will lead to peace and understanding. He writes, "Our new instant communication network may drive us apart emotionally just as it is drawing us together institutionally. Instead of new friends, we get new enemies."

Bombings of the USS *Cole,* our African embassies, 9/11, and threats from many Muslim groups underscore this point. National differences of religion, ideology, and geography are very deep. Whenever new intrusions or intruders violate our sense of time, place, and history—or even our self-interest—we resist them. We found this out when we "civilized" Native Americans. Their sense of resentment seems to be happening globally.

It's easy to change artifacts, difficult to change human nature. We say we will wire the world. How does this sound to the people, nations, and tribes being wired?

A YEAR OF SCANDAL. The Associated Press story about 2002 sums it up this way: Perp walks and handcuffs. Investigations and indictments. Look back at the highlights of 2002 and many of the images that come to mind suggest a police lineup.

Scandal topped the news in 2002. It swamped legendary companies, reshaped the political agenda, and triggered a crisis of credibility. A wide survey of U.S. newspaper and broadcast editors chose the implosion of WorldCom the year's biggest story, with the nation's limping economy a close second. Many top stories, from airline woes to interest rates, remained as the old year ended.

How big was the WorldCom scandal? It involved $11 billion in improper accounting, triggering the largest bankruptcy in U.S. history. Hundreds of employees were left jobless, without benefits. A chain reaction followed, involving rising unemployment and charges against many major corporations, such as Tyco, whose chairman's wildly expensive purchases included a $6,000 shower curtain.

Eleven interest cuts by the Federal Reserve didn't help much. The stock markets finished their third year of consecutive losses. This had not occurred in over six decades.

One fine quality of Americans, as the new century unfolds, is that they are still able to laugh at themselves and their problems. They know Mark Twain was right: human nature is widely distributed in the human race, and we can expect the unexpected. Many ups and downs occurred in 2003, but as the year ended, ups held the stage. In November the economy posted an 8.2 percent growth, and the Dow Jones Industrial average broke through 10,000. A new Medicare prescription benefit bill was passed by Congress, and holiday shopping and spending soared.

But the best news of all came by surprise on December 14, 2003. American troops found and arrested Saddam Hussein, hidden in a spider hole outside Tikrit. Coalition Administrator L. Paul Bremer announced the news that rocked the world: "We got him!"

New life and hope emerged as we moved forward into 2004. Old problems remained, but the twenty-first century held new hope and promise.

Notes from the Backbench

The outs and ins will always be with us. The ins sit up front, the proud majority; the outs, loyal opposition, go to the backbench. Democracies have always striven to protect outsiders who see and challenge things the insiders ignore. They are the watchdogs of democracy, as in the British Parliament. Many of our greatest leaders, such as William Pitt and Winston Churchill, have sat on the backbench, awaiting their term. Americans have had a continuous flow of backbenchers, including Thomas Jefferson, Abraham Lincoln, Teddy Roosevelt, and Franklin D. Roosevelt. At the moment long-dominant Democrats sit on the backbench in America. They are being heard, loud and clear.

Intellectuals and academicians have their backbench too. I discovered this when I entered Yale University, and chose to sit with professors such as Ralph Gabriel, Whitney Griswald, John Sirjamaki, and Norman Holmes Pearson. I had assumed I would join the front bench, the strong traditional disciplines, such as English, history, and the classics. Instead, I found new companions in American studies, sociology, and popular culture. They, alas, are gone now. But those of us who heard them still follow them. They taught their own outsiders who had come in: Walt Whitman, Henry David Thoreau, Mark Twain, Max Weber, Karl Marx, William Faulkner. They believed we should not shut doors but take off the hinges. We are still working at it. What a splendid time to be a backbencher!

So what if Americans are too fat, too self-centered, too chauvinistic, too addicted to bigness? We admit it. Take for example, the bloated SUVs that roam our highways and guzzle huge amounts of gas.

I have a June 1, 2003, pseudo news release from Detroit about the forthcoming Ford Exorbitant, which seats fifty comfortably, has a full kitchen, three bedrooms, one and a half bathrooms, and a Ford Explorer (the current super SUV) as a spare.

Why not give up living in your home and move into your Exorbitant, the announcement, put out by BBspot, suggests? There are numerous advantages. In emergencies, you can always use your spare Ford Explorer. Since your Exorbitant is nuclear, getting 70,000 miles per enriched uranium rod, all should go well. This is most reassuring. Isn't half a million dollars a small price to get peace of mind?

Theories of energy, matter, space, and time are in flux. What was once held as simple has become incredibly complex. Information fights disinformation, smut doubles with glut. Millions of terminals all over the world send messages with the speed of light. Each terminal has memory and editing power. Anything and everything goes. Data data everywhere and not a chance to think.

Communication is becoming compunication: the linking of computers, fiber optics, satellites, and the new toy of the month. Once all the world was a stage. Soon it will be a mall.

Who benefits from new technologies, mergers, buyouts? Will a new overclass usurp most of the power? Will those unfamiliar or hostile to the new electronic wizardry become a new underclass— undervalued, underemployed, and unemployable?

Revolutionary changes are always frightening, especially for those raised in the earlier print culture. We know how much earlier technologies—such as the automobile, television, nuclear power—raised grave concerns. We also know that some of those concerns were justified. The automobile, television, and nuclear power plants have raised problems—few if any anticipated or understood. The earlier wild euphoria has disappeared.

Surely we don't want to halt the flow of information, any more than we want to deny the many obvious advantages of technology. New "wonder drugs" are indeed wonderful; we have advantages undreamed of even a generation ago.

Is big better? Is biggest best? Who will arbitrate and regulate the Electronic Revolution? Who will protect our basic privacy? Is anything too slanderous, libelous, or pornographic to be excluded or censored? Do our honored doctrines of free speech and free press fit the electronic media? Where do we draw the line between the silly and the significant, the truth and the hype, the new and the neurotic?

As in the ancient parliamentary model, we are the loyal opposition, occupying the backbench, but expecting—even demanding—to be heard. If, as we believe, the world is out of joint, we must say so, and take our stand.

I'll Take My Stand is an important book published in 1930, by a group known as the Southern Agrarians. They believed and fought for a Southern regional tradition, heritage, and "way of life" which was under attack, largely by Americans who championed consumer capitalism and massification. The battlefield is global now; and those "taking their stand" now come from all corners of that globe. They still fear the consumer capitalism which would turn our wonderfully diverse planet into a standardized Global Village. They still fear a crass mass culture that champions conformity while praising diversity. They see the computer as the avant garde of dangerous trends, and refuse to let it dehumanize us. We are not ciphers. We are neither "0s" nor "1s" on the silicon chips and will not be computerized. Our world is neither black nor white but full of infinite numbers of grays. Postmodern psychobabble must not lure us into treacherous uncharted waters.

We intend to re-do old injunctions. Thinkers of the world unite. You have nothing to lose but your chips.

* * *

My goal is simple: to set minds working and tongues wagging. I think of popular culture as a topic of, by, and for the people—hence, of concern to everyone. Of course, everyone will not agree with what I say, or how I say it. That's where the wagging tongues come in. I hope my words will open up a dialogue, and that it will continue long after this book is closed or even forgotten.

It has many facets, like a diamond, and can well be subversive and explosive. Fun and games? Scorn may be mixed with the fun, and the games can be deadly serious. Popular culture is at the heart of revolutions. Great revolutions slip in on cat's feet. Those most affected by them, the elite and the mighty, seldom see them coming. Popular culture sees and hears, being close to the people. If the medium is the message, then the reaction is the revolution.

These are my thoughts as I look at a famous 1905 photograph which long bedecked my grandmother's parlor. The photograph, like my grandmother, is British. In it King Edward VII and Queen Alexandra pose placidly in Windsor Palace Garden. Edward wears a top hat and frock coat. Alexandra's posh hat is veiled and plumed. In the background an avenue of trees as orderly as the Coldstream

Guards stretches back to infinity. Behind them stretches the past, a sunlit corridor of order and stability.

The same year that photograph was taken, Albert Einstein wrote his paper proposing "the special theory of relativity," positing reality as a complex of masses and motion. Matter became shapeless, to be described only in abstract formulas. Matter and energy were conceived as distinct but interchangeable. Fusion, fission, and the atomic age would follow.

That fusion and fission also apply to popular culture. In it many different things blend and converge: the sound of laughter, anguish, ecstasy, pain; rockets ascending and atomic bombs exploding; babies gurgling and AIDS patients dying; the yells of tens of thousands at the Super Bowl, and the chirp of a single cricket in an unending Kansas wheat field. Put all these into your definition.

Be sure to note too that it thrives on opposites. It soothes and irritates, understates and exaggerates, inflates and deflates. Long on compression, it is short on compassion. That compression, epitomized today in the cartoon and the sound bite, has had many names over the ages: the quip retort, one-liner, epigram, bon mot. The list of the master compressors is impressive: Buddha, Aesop, Jesus, Paul, Petrarch, Pascal, Will Rogers, Jane Addams, and Rachel Carson, to name a few. And don't forget Winston Churchill, F. D. R., and John Kennedy. They all knew how to reach the hearts and minds of many.

With such giants to inspire and lead us, we can only hope to reach the hearts and minds of a new age.

PROBING THE POPULAR

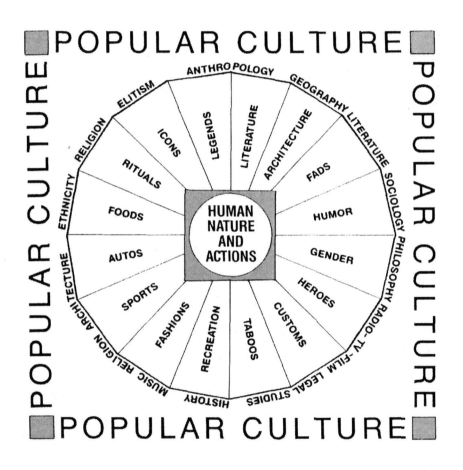

The Probing Process

"I think, therefore I am."

René Descartes,
Le Discours de la Methode

Let's start with the dictionary definition. "Probe—to investigate thoroughly; a tentative exploratory investigation." That sounds like a very good idea and mission. Why is it so difficult and controversial in the twenty-first century?

Information, disinformation, and speculation pour in from all sides. We suffer from information overload, and there is no cure. What can we believe, whom shall we trust, from the army of ad men and women, journalists, spin doctors, newscasters, publishers, manufacturers, merchants, and entertainers? There is a new Gresham's Law in our public life: counterfeit happenings drive genuine and spontaneous ones out of circulation.

The story of making these illusions, be they about O. J. Simpson, Madonna, Elvis, Bill Clinton, Al Gore, George W. Bush, or the Person of the Week, has become the most appealing news. The media are obsessed with creating and shattering "favorable images." In so doing, we may be demeaning and defeating ourselves. Our most dangerous enemy might not be racism, culture wars, ideology, disease, or demagoguery but the lure of illusion and unreality, of replacing ideals by images, aspiration by resignation. Cynicism sets in.

How can we digest, let alone evaluate, what confronts us on television, for example? There is good reason for calling it the boob tube. What about the spam that has polluted e-mail, or the endless ads that outweigh and outshine the news in newspapers? Ah, Plato, help us in our new glittering cave. Where does reality really begin?

Reality? What is actual, true, authentic, genuine? By whose definition? A soldier, philosopher, scientist, lawyer, novelist, artist, farmer might have quite different definitions. Are they all "true"?

Actuality? The instant between the ticks of the watch; a void slipping forever through time; the rupture between past and future; the space between events. Point zero; the voices of silence. The instant of actuality is all we ever know. The rest comes in signals relayed by outsiders. Can popular culture probe this? Where shall we begin?

How far back in the past shall we go? Should we follow the advice in *Alice in Wonderland,* and "Begin at the beginning"? That takes us far back beyond written history, into myth. All cultures develop a creation myth. Ours puts Adam and Eve in the Garden of Eden. Today we take another so-called "scientific" mythical road, involving split atoms, black holes, and the big bang.

We search for bones, send up spaceships, peer into telescopes, grope for dates. All this speculation is part of our popular culture. It is reflected in our comics, movies, TV, novels, and cartoons.

Startling discoveries emerge. We land a probe on Mars and unlock the DNA chain. In Africa we find a fossil of a young adult woman (who has been named Lucy) which dates back approximately three million years. Did popular culture begin when Lucy walked the earth looking for food, shelter, or a mate? Did she finally evolve over three million years into Lucy Locket or Lucille Ball?

We can trace our roots back to the Neolithic Age, when agriculture and tribal life developed. We know little about the individuals who lived then, in millennia and centuries when history was blind. Much later the world was peopled with whole armies of serfs, slaves, prisoners, and peasants, unlettered and unknown, buried in history's darkest dungeons.

Known or unknown, they were part of popular culture. Most have no memorial. Still, they were linked to use by unbreakable chains. Popular culture is also common culture. Even the origin of the word *common* is unknown.

What is known is that remarkably successful cultures developed around the Mediterranean, of which the Egyptian, Greek, and Roman are best known and documented. America is often called "the new Rome," with an empire not unlike Rome's built on economic, military, and political power as we enter a new century. Like Rome, we exploit the media (plural of the Latin word *medium*—intermediate or between).

The Romans used Latin as the predominant medium in their empire; as English is being used in ours. Media make power, information, cooperation, and popular culture possible.

After the fall of Rome, the Holy Roman Empire, which lasted a thousand years, was the wellspring of Western power and popular culture. What has replaced it?

The tool of survival is tradition: the handing down of information, beliefs, and customs from one generation to another without written instruction. This provides continuity and makes daily life meaningful. Tradition follows two separate tracks: the Great Tradition and the Little Tradition. The former sustains the elite, who attend schools and universities, seize or inherit power, set styles, and direct affairs. The Little Tradition includes all the rest: the nonelite, unlettered underclasses.

Those outside the Great Tradition quite literally "didn't speak the language," which in the church, courts, and universities was Latin. The Roman Catholic Church used a Latin mass well into the twentieth century. The educated elite, having access to both traditions, were usually bilingual and bicultural. The Little Tradition found its roots in fields, stables, taverns, kitchens, and marketplaces. There it preserved folktales, folksongs, mystery plays, fairs, and festivals, taken together, popular culture. Its great days were Christmas, carnivals, May Day, and Easter. Some of those days' greatness still remains.

The Little Tradition lived on with peons and peasants; shepherds and sailors; minstrels and magicians; beggars and thieves. Our best accounts of all this are Peter Burke's *Popular Culture in Early Modern Europe,* and Robert Redfield's *Peasant Society and Culture.*[1] These books, augmented by *Funk and Wagnall's Standard Dictionary of Folklore, Mythology, and Legend,* help us understand the preindustrial world, and to re-enter the world we have lost. We may even come to admire the people from whom our own culture has sprung, and how popular culture is transmitted.[2]

The common (or popular) culture of the preindustrial world is ephemeral and elusive. Hence it is important that we try to re-create this vanished world.

Demos is the Greek word for people. *All* the people? Hardly that. The Helots in ancient Sparta were nonpeople, with no rights or political voice. Democracy (*demos,* the people + *kratia,* authority) was, said Plato in *The Republic,* mob rule, the worst possible form of government. We are startled, taught from childhood that democracy is the best form of government. Doesn't our Constitution begin with the famous words "We the People of the United States"? We seldom ask just how many of "the people" were consulted when the Constitution was written or were allowed to vote when it became the law of the land. No one intended *all* the people to vote. They never did and they still don't.

The *demos* have a long and complex history, beginning centuries before the Greeks and their *demos*. That history continues into modern times, and into our new millennium. Here we follow the pattern of the ages. Common culture and people were for thousands of years largely illiterate, and words fly by. The little we know about all those people comes from the words of literate outsiders. Such outsiders did and do find it difficult to understand people unlike themselves.

In our new century we must pay more attention to the early roots of popular culture in other nations, and increase international studies. One major pioneer was Japan's Hidetoshi Kato, whose *Japanese Popular Culture* (1959)[3] pointed out that his country became the first to form a mass audience. Popular culture came into maturity when the nation's surplus energy went into cultural activities and reading.

Popular culture thrives on new ideas and serendipity. Chance and surprise discoveries have played a major role in history. The Trojan Horse defeated Troy. Japan's surprise attack on Pearl Harbor changed the course of history. The Battle of the Bulge in World War II, the Cuban Missile Crisis, the 9/11 attack on New York and Washington all changed our policies, fears, prospects, and level of fear. All this affects our popular culture and our technology. These things go bump in the night.

By 2003 more than two million hidden security cameras were peering down on America from countless sources. That's a lot of peering. We depend more and more on e-mail, which might include our most intimate secrets, without recalling that it can be retrieved, even after it's deleted.

On the Internet, Web site, store receipts, credit card receipts, we leave our electronic footprints. A clever snooper or enemy can uncover our tastes, choices, Social Security number, even some of our intimate thoughts, without our knowing it. Surely this isn't what our Founding Fathers envisioned or expected.

Much of my material is about recent centuries, and Western civilization. There is a central theme: the parameters of popular culture need to be expanded and explored. I reject those who think popular culture is a newcomer to the human drama, that it is only electronic, or that it sprang up in the 1960s with the beatniks, hippies, new journalists, pop artists, and celebrities such as Andy Warhol, Elvis, and Marilyn Monroe.

Popular culture is as old as humanity itself. Culture has always been "popular," thriving on formulas, archetypes and stereotypes, fads and follies. It is circular, repetitive, and powerful. Here today and gone tomorrow—but just wait, and it will be back next year, next decade, or even next millennium.

We like to think of ourselves as inventing mass production. Yet the Sumerians were mass producing funeral effigies centuries ago; so were the Egyptians, Persians, Incas, Tibetans, and Chinese.

We may be in a new space age—but we are also neo-Victorians in many aspects of our lives. The French had it right: *plus ça change, plus c'est la même chose.* So why should we be scornful of ephemeral popular culture?

The situation is complicated by having to produce endless music, programs, movies, ads for use 24/7, around the world. David Brinkley, one of the best major newscasters, refused to increase his prize-winning program from half an hour to a full hour. "There just isn't that much news," he insisted. So how can networks such as CNN and others grind it out twenty-four hours a day?

In our electronic age, mass production can flash simultaneously around the world. If there's a fine new tennis ball in America, Australia will know about it in hours. If Saddam Hussein issues a warning, it will be not only for America, but all the world, soon after he delivers it. Popular culture is the message and the whole world is the receiver.

All this makes our probe not only difficult, but in one sense indeterminate. How do we know how the message hits billions of eyes and ears almost simultaneously? What will make an impact? What will drift off into cyberspace? Who wants to read yesterday's newspaper, or hear the speech that's already been featured a dozen times on television or the Internet?

So we end up with paradoxes and contradictions rather than clear-cut definitions. We have looked for common denominators, but ended up with pat phrases and clichés. We are fascinated with and demand change. It is a mirror held up to the present, but drawing from the past, full of motion and madness. It laughs at the bureaucrats and revels in the wonder and whimsy of life. All definitions turn out to be inadequate.

American culture, on various levels, is process more than product; the process of motion across oceans and mountains, going west and coming east, moving up and down the social ladder. What is basic to our popular culture is this obsession with process as reflected in the work of our heads, hearts, and hands.

We are broadbacked and boisterous, with Paul Bunyan's strength, Casey Jones' daring, and Salvation Sal's devotion; John Henry's rhythm, Louis Armstrong's jazz, and Huck Finn's charm. We build tense turbulent cities, with fingers of power, skyscrapers, reaching out into space; stately plantations warmed by glowing memories; earth richer than the treasures of Solomon; launching pads to send astronauts to the moon.

This is a land of constant surprises, imitation castles, glass domes, CEOs, sports mania, freedom marches, little magazines, big sells, wetbacks, flashbacks, comebacks. We may all be famous for fifteen minutes.

Look at the names on the land: Dry Bones, Nantucket, Go-to-Hell Gulch, Lost Mule Flat, Lake June in Winter, Machopongo, Bubbleup, Wounded Knee, Purgatory Creek, Poor Man's Hollow, Okaloacoochee Slough, and Boondocks. Ours is a continent of sharp bright cold, long Indian summers, and sapping heat. Our two-billion-acre farm is ideal to grow corn, wheat, weeds, powerful men and women, dissent, and rebellion.

How to cope with such a land?[4] There is no one way, perhaps not even any combination of ways. And just where does popular culture fit in? Instead of definitions, we might try metaphors.

I like to think of popular culture as a wandering minstrel, a thing of shreds and patches, a mercurial mosaic that changes as we watch it. Like Proteus in Greek mythology, it can take on forms and shapes of all things and all places. We turn now to our pioneer probers, who led the way in studying and exploring popular culture.

Pioneer Probers

Unscrew the locks from the doors.
Unscrew the doors from their jambs.

Walt Whitman
Song of Myself

History abounds with pioneers who thought and studied about what we now label *popular culture*. They came from many times, places, and cultures. We shall mention only a few before commenting on the American pioneers.

Pivotal thinkers of China, India, and Egypt come to mind. Both the Old and New Testaments should be included and are "popular" today. Greek philosophers such as Thales, Socrates, and Aristotle must surely be included as well as the Roman Empire, built around a common language (Latin), law, and rituals. Romans used the media (plural of the Latin word medium, intermediate or between) to great advantage.

No one comprehended this better than Marcus Tullius Cicero (106-43 B.C.) who helped make Roman language and customs universal and added important new words and concepts. He replanted the Greek way of thinking, the dialectical, and championed the academic way of life. If popular culture may be said to have a godfather, he is it.

After Rome fell in the fifth century A.D., the Great Tradition of empire and progress ended and the Little Tradition prevailed—folk tales, songs, mystery plays, fairs, and festivals reflected the popular imagination. Eventually power was consolidated and the Holy Roman Empire emerged under Charlemagne, who was crowned in Rome in 800 A.D. It would last a thousand years. New nations, alliances, and centers of learning sprang up, and what we call the modern age began; "the people" would finally find their voice and their champions.

Able European scholars and historians strove to preserve and study what they could of premodern culture: Thomas Percy, Samuel Johnson, J. G. von Herder, Sir Walter Scott, and the Grimm brothers, for example. They collected popular songs, legends, and folk tales, transmitting priceless material. We met Everyman full-blown, when John Bunyan (1628-1688) published *Pilgrim's Progress* in 1678. By then the first pilgrims had reached American's shores, and the progress would continue in the new environment. The long trail from Imperial Rome to Imperial America still invites much more work and probing.

Striving to found New Canaans and New Havens, the early English pilgrims concentrated on religious literature and sermons. They carried European manners and mores across the Atlantic, and waged wars against the natives whom they called Indians. The term *popular culture* would not appear for many generations. The "people" would clear forests, found cities, and dream of new "cities on the hill."

Gradually a new literary tradition began, centered in New England around such figures as Emerson, Thoreau, Longfellow, and Whittier. Their contributions are well-known and studied. Benjamin Franklin, centered in Philadelphia, gave us *Poor Richard's Almanac,* supplying many of the shrewd insights and practical advice that have guided the newly united states ever since.

But the major nineteenth-century pioneer was Walt Whitman, whose *Leaves of Grass* is part of the canon of all levels of culture. He "sang the body electric," rode on the Broadway horse car, served as a male nurse during the Civil War, and gloried in the street life of his beloved Manhattan. In his poem "By Blue Ontario's Shore" (1855) he asked two probing questions: "Are you really of the whole People? . . . Have you vivified yourself from the maternity of these States?" Though he never used the term *popular culture,* Whitman was one of its best chroniclers, who finally came to personify it.

The pioneering work of Charles Beard, Gilbert Seldes, V. L. Parrington, Van Wyck Brooks, Randolph Bourne, Ralph Gabriel, and Lewis Mumford appeared in the 1920s and 1930s. But it was World War II, belief in "the American Century," and America's phenomenal success that forged a keen belief in American distinctiveness. If only we could explain this to the old world, they might become partners with our democratic new world. American scholars rose to the challenge. In 1948, University of Minnesota's Professor Tremaine McDowell formulated his "First Law of American Studies": Present the complex design of American life so as to reveal the fundamental diversity of human experience, hoping we might eventually find a fundamental unity. He and enthusiastic colleagues founded the *American Quarterly* in 1949, and the American Studies Association in 1951. This gave the movement the stamp of professional approval and status. Some were skeptical about this new enterprise.

Meanwhile new programs sprang up around the nation; the dollar reigned supreme, and the United States Information Service (USIS) was well funded by Congress. The formation of the European Association for American Studies in 1954 gave the movement a new status, as programs also sprang up in Latin America, Africa, Australia, and Asia.

All this activity helped advance the cause of popular culture, even if it did not always get equal billing. Those of us who were in the field discovered that people not only in the United States but around the world were fascinated by our music, films, comics, and lifestyle. The door was open for what came to be "The American Century."

Talk of America's new role tends to invigorate its scholars and calls upon everyone to assume his or her part. The implication was that we should not only explore but also defend our democratic way of life.

At Yale University, where I came from the U.S. Navy in 1946 to the American Studies graduate program, President Charles Seymour announced its goal: "To produce a solidified faith in America and devotion to its maintenance." Many writers and scholars took a pro-American stand on the Cold War and received widespread support for doing so. There was also a heavy stress on "the Atlantic community" as the best framework for investigation, which became a kind of academic NATO (North Atlantic Treaty Organization). Some of us who entered the field might have become, without being fully aware of it, Cold War warriors.

This was not the only impetus. Early enthusiasts had a more universal aim: to define a national mind through a new interdisciplinary approach. They saw this as a kind of crusade. Hence the movement would take on moral as well as professional dimensions. Professor Ralph Gabriel, my mentor at Yale University, liked to cite the book anthropologist Ruth Benedict wrote during World War II at the government's request: *The Chrysanthemum and the Sword*. We wanted to know whether, in the event that Emperor Hirohito asked the Japanese people to surrender, they would do so. She predicted that, after considerable internal struggle, they would surrender. Gabriel suggested that this pointed to the possibility of meaningful studies of complex cultures.

Not all Europeans turned their backs on the *volk*. Gottfried von Herder (1744-1803) championed the idea of *Volkgeist*—the spirit of the people. It sprang, he firmly believed, from tradition, environment, and the times. He collected scores of *Volkslieder*—songs of the people—and *Volksmarchen*—folk tales. This idea spread. The pioneering Herder also believed that nature and history obey a uniform system of laws. Incredibly prolific, thirty-three volumes of Herder's work were published between 1805 and 1820.

In a series of letters and articles, he raised crucial questions: How can we begin to understand people who neither read nor wrote and left no records? Were the elite informants able to "get into" peasants' minds, relate to their feelings? Might they have been misled or deceived as in the famous twentieth-century case of anthropologist Margaret Mead in her highly touted book *Coming of Age in Samoa*?

If we have a single text for an important topic—such as seventeenth-century Russian tales and folk songs, collected by two British visitors (Richard James and Samuel Collins)—can we assume the collection is "authentic"? What about popular events known only through "the authorities" who suppressed them? We only learn about heretics and witches, for example, through their prosecutors. People burned at the stake or hanged do not get to give their rebuttals.

Can we really understand active performance (such as festivals, dramas, or sermons) if we have only words, not actions, to guide us? Who chooses those words?

Still, the European mind-set continued to guide America. Even our Declaration of Independence, which held that all men are created equal, drew heavily on British thinking, especially that of John Locke and Edmund Burke. In 1829 Emerson complained we had listened too long to the "stately muses of Europe," but we did then, and we still do. Well into the twentieth century the works of T. S. Eliot, Ortega y Gasset, and French existentialists, echoed by Americans such as Ezra Pound and Dwight Macdonald, had only scorn for what became known as popular culture. They claimed it perverted high culture into something "quite unnatural." It was trivial, repetitious, obnoxious, even obscene.

How things have changed! The culture they scorned has become the lingua franca of the evolving electronic Global Village. European artists pioneered in areas intellectuals and writers scorned. Dada, futurism, surrealism, and art nouveau helped set the stage for pop. F. T. Marinetti's *Futuristic Manifesto* (1909) sang the praise of speed, machines, and motion that were featured in popular culture. They helped make it possible for America to produce a Walt Disney. Born in the last year of the nineteenth century, Disney invented Mickey Mouse en route to Hollywood in 1927. Mickey championed pop style, and won the heart of the world, under such different names as Michel Souris, Miguel Ratoncito, Miki Kuchi, and Mikki Hirri. Later on he would be the featured star in Disneyland and Disney World, as well as in theme parks opening in France and Japan.

Critics with a new orientation stepped forward in the early twentieth century. In 1915 Van Wyck Brooks claimed in *America's Coming of Age,* we must learn to understand three "brows": highbrow, midbrow, and lowbrow. Gilbert Seldes' groundbreaking book *The Seven Lively Arts* (1924) set out to restore a good conscience to people who enjoyed circuses, silent movies, and comics. John Kouwenhoven and Sigfried Giedion praised the unselfconscious effort of common people to create new satisfying patterns, inspired by the driving energies of America.

That driving energy would emerge in World War II, and would change views and attitudes about America forever. It would also give new meaning to popular culture. The term appeared for the first time in 1960 in the *Reader's Guide.* Since then it has become omnipresent, especially in the media. Under various guises and titles, it is taught at scores, if not hundreds, of colleges and universities.

The Popular Culture Association was founded in 1967, leading to the first national meeting in 1968. A journal and a press, the Popular Press, were founded in that same year, and have flourished ever since. As a result, large numbers of papers were presented; friendships were begun; and people who had tired of worn-out topics and disciplines joined.

The new organization served scholars by providing a way of incorporating what others considered a "vast ooze of the mass mind" from folk festivals to movies, comics, and sports as legitimate subjects of study. Students were attracted to subjects made "relevant" by the focus on their own world as teachers applied tools such as icon, genre, and motif to the range of popular phenomena. The movement challenged the "new critical" orientation which was dominant in much of the humanities. Since popular objects were shaped by market forces in their inspiration, production, distribution, and consumption, it was difficult to hold to the new criticism dogma that the work itself as read, seen, or heard in the abstract was the only legitimate subject. The popular culture orientation put the artifacts in social context and challenged the assumption that intellectual discourse could take place abstracted from life.

Ultimately what is involved is the democratizing of scholarship. No one said this would be easy: many think it impossible. Those intellectual aristocrats who thought so much and so well managed for centuries when the connections between their thoughts and the sources of power were direct. When those trained by and loyal to the church chose popes, and those who wore the old school tie ran cabinet, college, and empire, elite methods really worked. That day has passed.

A combination of factors still not fully understood is changing all this: technology, ideology, and mobility. Instead of being a series of propositions for professors to debate, Marxism has become a moving force in hundreds of millions of lives. The power of the rifle is not lessened if the man wielding it is not properly educated or dressed. There is no way of segregating the electronic beeps that have turned the atmosphere into a source of constant revolution.

Even the word *revolution* does not and never will again mean what it meant when rocks, guns, and bombs could decide issues. Not only on the battlefield but in the media will the battles for men's minds be won or lost. But media history is as long as military history. To understand, we must not only look around, we must look back.

Foreign countries such as Japan, Britain, and Germany led the way, as popular culture became the norm, and mode, for many young people. Commenting on Germany's International Book Fair in 1970, Peter Härtling wrote, "This has been a fair of pop singers, writers, and enthusiasts."

Both the 2000 millennium and the 2001 World Trade Center attacks provided new material and interest for popular culture, as did global film festivals and youth rallies. One estimate claims that a million students now study popular culture in universities. The strict canon has become much more fluid, not only in America, but also around the world.

The new opening was documented in a book published by the Popular Press in 1999 called *Pioneers in Popular Culture Studies,* edited by Ray B. Browne and Michael T. Marsden. They thought of pioneers as brave souls who move into uncharted territory, meet unpredictable dangers, and face the hostility of "established" disciplines and subjects. Hostility was especially strong from journals and universities where the "new criticism" reigned. Depending on close, intense reading, these critics scorned or ignored accessible and popular literature.

But these elitists were isolated from mainstream thinking, both at home and abroad. In a democracy, as new pioneers and probers firmly believed, the real vitality dwells in the burgeoning life around us. The new media made this ever more apparent. Marshall McLuhan proved to be right: the medium became the message and the massage.

The development of popular culture was advanced by new activity and success among women and long-ignored minority groups. Thanks to them, we are greatly broadening our scope and scholarship, opening doors that have been closed far too long.

In a democracy, even a developing democracy, all institutions and citizens need to be studied, and their activities documented and preserved.

Probing a Frenzy

"We're Number One!"

In America we are much prone to popular culture frenzies. These things come like tornadoes on land, hurricanes at sea. They catch us unaware, quickly change the landscape or seascape. We are caught up, almost overwhelmed. How long will it last? What does it mean?

Consider the frenzy which occurred on July 10, 1999, when we went "down to the wire" in a soccer match, won a victory on the basis of penalty kicks, and emerged world champions in the Women's World Cup Soccer finals. The immediate popular slogans were like Emerson writing about the American Revolution with his "shot heard round the world":

YES I CAN!
U.S. WOMEN RULE THE WORLD!
WORLD CUP FIELDS NEW ROLE MODELS! ·
READY TO FOLLOW THE BOUNCING BALL!

This breathtaking victory took place to a sellout audience of 90,185 in Pasadena's Rose Bowl. World crises were on hold in the White House. Both President Bill Clinton and First Lady Hillary Rodham Clinton were there to cheer on "our team." Not many winning teams have had that kind of political or photo-op support. Forefingers went up ("We're Number One!") all over the land. Winner take all.

But the main photo-op came when Brandi Chastain tore off her shirt to reveal her new soccer bra to the watching world. Having bared herself, she clenched both fists and opened her mouth so wide she could have swallowed a lemon. That was the picture flashed around the world, ending up on the cover of *Sports Illustrated, Newsweek,* and many other magazines and newspapers. A new superstar (already nicknamed Holly Wood) was born.

All this was not, however, spontaneous. Fifteen months earlier, Brandi Chastain had worked out a deal with Nike, explaining what she wanted. Ann Gerhard, a *Washington Post* reporter, found a crucial Brandi quote: "If I know that at some point I might take my shirt off, I'd like my bra to look nice. I'd like someone to look at me and go, 'Oh that's nice—I'd wear that!'"

She had posed once before in what she called the "altogether" (nude?) using a well-placed soccer ball. She joked that it only took ten minutes of being "in the altogether" in front of a camera before she relaxed.

But bra makers didn't relax after the stadium unveiling. Nike, gloating over the payday, rushed a new line into production and arranged promotional appearances. Nike competitors, such as Reebok and Fila, estimate Chastain's seminudity will do wonders for the $230-million-a-year market. They expected "exponential-type growth." Chastain already had a prior contract, said to be "in the five figures," but her agent announced it should be renegotiated for much more.

Sports shoe companies, soccer equipment firms, TV talk shows, cereal box decorators—all rushed to the starting gate, and so did the politicians. The Clintons led the parade, inviting the whole team to the White House with the acclaim usually reserved for returning successful four-star generals.

Statistics of the U.S. World Cup Soccer victory, which the United States had won in 1990 without much fanfare, and lost in 1995, are impressive. In 1999, 650,000 tickets were purchased for the thirty-two matches in the United States, bringing in $23 million. On July 4, when the U.S. team played Brazil, 2.9 million households watched. Over 100,000 girls began playing soccer between 1990 and 1997. This caught the attention of Washington spin doctors, who courted "soccer moms" in the elections. No one knows how many there were or how many soccer moms (as with many of my friends) were actually soccer dads.

All this is the climax of the growing attention to all aspects of athletics. They breed frenzies. Will Michael Jordan come back and play basketball in 2004? Will the baseball players (who often are millionaires) strike over contracts in that same year? Other American sports stars had shone but faded in past years. Most sports have been traditionally male dominated, but that is changing rapidly, greatly increasing the frenzy game.

In 1965 Donna de Varona became the first female TV network sportscaster. In 1972 Congress passed Title IX, banning sex discrimination in federally funded sports programs. That opened wide the gates. Fifty million people watched Billie Jean King beat bigmouth Bobby Riggs in the 1973 "Battle of the Sexes" in tennis. Three U.S. female teams won gold medals in the 1996 Olympics.

Team sports grew. The Women's National Basketball Association averaged 11,000 fans per game in 1998, and the number rose 12 percent in one year. The era of skirted women skaters, such as Dorothy Hamill and Michelle Kwan, gave way to tiny teen tumblers such as Mary Lou Retton and Kerri Strug. Now the spotlight favors lean muscular women who grimace and score. Female high-fives, hugs, and yells are in. Two of the reigning stars are twins Venus and Serena Williams, known for their tennis skills and iconic hairstyles. They are already swimming in endorsement deals, as are numerous male athletes. Then came the 1999 female soccer team frenzy and popular culture took over.

A week after the ecstatic outburst, the soccer players' attention turned from celebrating to negotiating. Endorsement royalties will be peanuts compared to contracts underway to start a new Women's Professional Soccer League and a Scheduled Indoor Tour by "America's Soccer Team" (tag invented by Bill Clinton) or "The Girls of Summer" (coined by Hillary Clinton). The proposed international tour would include, for a start, Egypt, South Africa, and Australia. Journalists noted that the 2000 Olympics would be held in Australia. The tour would certainly warm up the waters.

Here we move into multimillion-dollar deals. Unfortunately the grandiose plans drew a question mark from U.S. Soccer Federation President Robert Contiguglia. Schedule conflicts might result. Nevertheless, the alert managers arranged for the female stars to play eleven matches between October 22 and December 1, 1999, in U.S. arenas nationwide. SFX Sports Group, organizer of the tour, guaranteed the players $50,000 apiece, win or lose. Meanwhile the team, anxious to stay in the news, presented an autographed soccer ball to vice president Al Gore and his wife Tipper. Clinton got his own U.S. jersey ("Clinton 99") and wife Hillary a crystal soccer player statuette. When you're hot, you're hot.

Just what does all of this mean coming in the middle of a period of national mourning for John F. Kennedy Jr. and two female companions who died in an airplane crash? It speaks to the nature of popular-ity, and fast-moving media hype. The very nature of popular culture is that we try hard to "Go and catch a falling star," to use John Donne's famous phrase. This is no new thing.

We remember similar frenzies over little Jessica trapped in the well, O. J. escaping in his Bronco, Tonya Harding agonizing over her broken skate laces, and Britain in trauma over Princess Diana, dying in a speeding Mercedes in a French tunnel. These "stars" have al-ready faded. Today's young students stare in disbelief. They have a whole new set of idols and wannabes.

Life is long, fame is short. Old timers try to remember. It isn't al-ways easy. "I'll never forget dear old what's-his-or-her-name . . ."

Keep your eyes on the heavens. Can you see the new star? How long will he or she shine? Weep not; others will soon follow.

There are countless frenzies to probe. We must run fast just to keep up. Our closets and attics are full of things put there by planned obso-lescence. Recall T. S. Eliot's pungent line: "Hurry up, please; it's time."

Don't try to "keep up" with popular culture. You can try, but can you do it? Like quicksilver, it will slip from your hands. Marshall McLuhan said: "If it works, it's obsolete." How quickly the "in" is "out," and today's superstar is *who?*

I find ample evidence in my own life. Childhood, in the 1930s, was shooting marbles in the backyard, playing touch football in a vacant lot, "kicking the can" in the street, avoiding the occasional automo-bile, going to the dime Saturday morning movie, invariably a western, featuring such paragons as Hoot Gibson, Ken Maynard, and Buck Jones, and on lucky Saturdays, maybe even seeing an animated car-toon, with a spindly Mickey Mouse who might appear as Tugboat Willie.

Then came World War II with food rationing and gas rationing, bond rallies, D-Day, and finally the ecstatic celebrations of victory at home and abroad. America, long on the backbench, suddenly moved up and became Prime Minister of the World. We didn't have toy war games, we had experienced a real war.

The 1950s brought the GI bill, gas-guzzlers with fin-tails, new western heroes (Davy Crockett and Roy Rogers) and out-of-the-world newcomers such as Commando Cody, Roger the Robot, and Space Cadet. My sisters, meanwhile, got a doll unlike any they had seen before: Barbie. There were exciting gender-neutral toys too, such as Wham-Os, hula hoops, and Frisbees. Most fads were pre-electronic, but some kiddie record players were around, and so were skateboards and Mr. Potato Head.

G.I. Joe, the world's first "action figure," made the scene in the 1960s along with considerable action with civil rights, segregation, and the new hippie lifestyle. Our world seemed to turn upside down when President Kennedy was assassinated in 1963, but there was still plenty of fantasy: *Mad* magazine, Monster Magnet, remote control cars, Sno-cone machines, and Zillion Bubble Blower.

Stretch Armstrong was big in the 1970s, fusing "tug of war" with the action figure. Along came Bionic Woman, KISS dolls, Evel Knievel, Hungry Hungry Hippos, and all the trappings that went with Star Wars and Planet of the Apes.

Home video games made a big comeback in the 1980s with Nintendo, while Teenage Mutant Ninja Turtles were everywhere, along with Cabbage Patch Kids. There were muscular He-Man and his nemesis, the evil Skeletor, and a new, smaller version of G.I. Joe. Do you remember the Care Bears, Gloworm, and Dungeons and Dragons?

In the 1990s times were good and toys expensive. Tickle Me Elmo and Beanie Babies cost a lot, and Furbys took us into high tech. Video games flourished, with entries such as Sega, Sony, and Nintendo. The digital age got stronger. Prices for toys went sky-high. So did our space ships. With a new century came new games, new fads, new prices. Go to the stores and find out for yourself. Go and catch a falling star.

Time passes; faces change, frenzy finds new faces. In the new century, dominated by gender wars, frenzy often centers on women, especially in sports. In May 2003, Annika Sorenstam was highlighted as "the world's best woman golfer," the first woman in fifty-eight years to compete in the prestigious PGA Tour in Fort Worth, Texas, where for years top male golfers had the course to themselves.

Her appearance was electrifying and started a media blitz. Major networks and news magazines featured her; she was on the front cover of leading tabloids. Crowds flocked to the golf course. Those who couldn't get in hung on the fence outside. Her every swing became a TV special. The featured photograph in most outlets had her with arms over her head, her fists clenched, and tigerlike determination on her face. She was "in." She was instant celebrity—the right person at the right place at the right time. President George Bush announced that he was "for her." Annika was a popular frenzy—even an idol.

The first day on the greens went well. She held her own in the Colonial golf tournament and ended in the middle of the pack.

"It was more than I could have expected," Sorenstam told the wildly enthusiastic crowd and press. "It's just a thrill to be here and I am very, very pleased with the way I played."

The second day did not go so well. She missed three putts, and being out of the top seventy, was cut from the roster.

Some of the sound and fury quickly subsided. In a culture where "winner takes all," she was out of the running. But in other ways she had won, as all the sponsors and publicists of the tournament knew. She had breached the gender gap.

As with most other sudden demigods, her reign was short. Two weeks later some academic feminists in my university were arguing that Sorenstam had made a permanent dent on a male-dominated culture and changed America's outlook. One of them proclaimed: "I'll never forget Ms. What's her name?"

Perhaps that says it all.

Hype haunts horses as well as humans. Consider Funny Cide, the "miracle horse" of 2003, who captured two-thirds of the Triple Crown (Kentucky Derby and Preakness) and drew 101,864 rabid fans to Elmont, New York, expecting to see his third victory, thus becoming the first Triple Crown winner since 1978.

All signs were go. Belmont Park was his home track. His past five workouts on the track had been very fast. Funny Cide was going to pound his five rivals to death.

The media had played it to the hilt. Funny Cide was on every major newscast, talk show, and front page. This was a "rags to riches" story, the kind popular culture loves.

All the good feelings soon vanished. Funny Cide finished third. The national uproar about the modestly bred, cheaply purchased overachiever was over. Rita Santos, wife of Funny Cide's jockey, Jose Santos, and her eight-year-old son collapsed tearfully into each other's arms.

Rain-soaked fans filed out silently. They had come to see a blue-collar New York horse, owned by a group of regular guys, grab greatness. Another frenzy had been buried in the mud.

American professors and authors have carried American ideas and books around the world. This picture shows the library in the American Studies Centre at Hyberabad, India. This is the largest American library outside the continental United States. (*Source:* Courtesy of Matt Tamagni.)

The American Studies Link

Popular Culture Studies is in some ways a stepchild, or break-off, of American Studies. Pioneer American Studies, usually combining literature and history, appeared in the 1930s, but the major expansion both in the United States and worldwide followed World War II.

Americans soon learned that this new fascination with Americana could serve us well, and a federally funded Fulbright Program sent hundreds of Americans abroad to explain and justify America. I was one of those grantees, and one of the many who later found other outlets and opportunities—such as the Popular Culture Association—that better served our needs.

American Studies was never unified or cohesive, trying somehow to mean all things to all people. Some distinguished scholars and writers gave it credence: Whitney Griswold and Ralph Gabriel at Yale; F. O. Matthiesson at Harvard; Tremaine McDowell at Minnesota; Willard Thorp at Princeton. Still it was always several disciplines in search of a methodology, a loose confederation that depended more on personalities than principles. The shifting emphasis from linguistic and belletristic analysis to cultural and intellectual history was a key motivation, the pioneer work of men such as Parrington, Brooks, Mumford, Lovejoy, Babbitt, and Lowes was stimulating. So was the electrifying interdisciplinary scholarship of the Progressive Era. From 1910 until the late 1940s, Irwin Unger notes, the giants of the historical profession devoted their scholarly energies to exhibiting the long and honorable record of insurgency and dissent on the frontier, on the farms, in the mining camps and factory towns.[1] Some of that energy powered the first American Studies circuits.

The critical impetus, beyond all doubt, was World War II. "Discrepancy between the position of the United States in the world and its place in curricula had long been growing," Sigmund Skard writes. "After 1945 it proved intolerable. A radical change was brought on by the need for reorientation."[2]

Looking back to the Truman era, when only the United States had the atom bomb, and Pax Americana was (like the mythic prosperity for which FDR struggled so valiantly) "just around the corner," we see a strong self-congratulatory note evidenced in all we did and said. The case for America's phenomenal success rested on the doctrine of Americanism—our melting pot, frontier, economy, leadership.

If only we could explain this to the world (especially Old Europe, tired to the point of exhaustion) they too would understand, model themselves on us, and become partners with the free (and brave) New World. Thus did circumstance and chauvinism motivate the American studies movement.[3]

The overseas response was dramatic.[4] Of course, we paid the bills for our Fulbright scholars, who suddenly turned up all over the world, for the United States Information Service (USIS), the Central Intelligence Agency (CIA), Voice of America, and Radio Free Europe, all of which deluged the world's libraries and air waves, and for the Marshall Plan, as well as tanks and planes providing the punch to our argument. Still the world *did* listen, Europe *did* recover, we *did* champion the new nationalism in Asia and Africa. No one will ever know whether American Studies abroad would have taken root without such massive support. Everyone knows that the support was provided, and the plants blossomed.

The movement's crest came at midcentury when the American Studies Association (with its official publication *American Quarterly*) was founded, and the multivolume *Literary History of the United States* (edited by Spiller, Thorp, Johnson, and Canby, founding fathers all) appeared. In that same decade Sigmund Skard's impressive *American Studies in Europe* and Robert Walker's *American Studies in the United States* appeared. But the growth pattern they discovered, and the optimism they reported, did not continue into the 1960s. Significantly, Professor Walker himself left the academic profession to join the federal bureaucracy.

Professor Walter Johnson's 1963 report for the U.S. Advisory Commission on International Educational and Cultural Affairs (published as separate brochure titled *American Studies Abroad*) reflected the new climate. It argued for sending abroad scholars "well trained and established in the individual academic disciplines," rather than men with interdisciplinary training and inclination. Johnson argued, in effect, *against* the whole interdisciplinary enterprise called American Studies. So did influential scholars throughout the land.

Other critics were far more outspoken than Johnson. Alfred Kazin, a literary scholar whose volume *On Native Grounds* had been considered an interdisciplinary achievement, spearheaded an attack on what he called the "collective autoanalysis" involved in American Studies which sanctioned "Any borrowing from sociology and esthetics, any enforced joining of the unjoinable; any political nostalgia or irritable political reflex."[5]

Who would answer or refute him? The movement's only "official" voice, the American Studies office housed at the University of Pennsylvania, was staffed only by a secretary and several teachers released from part of their teaching duties. Executive committee members, scattered throughout the country, fully employed in other tasks, met only occasionally. Until the fall of 1967, there had never been a single independent American Studies Association meeting on a national scale.

There were, of course, impressive regional programs, and splendid individual achievements, but no essential unity of purpose or program. An even more serious weakness had been exposed by Henry Nash Smith (the senior American Studies PhD holder in the country) in a 1958 essay called "Can 'American Studies' Develop a Method?" Smith concluded that there was no ready-made method in sight:

> We shall have to develop one for ourselves, and I am afraid that at present we shall have to be content with a very modest program. The best thing we can do is conceive of American Studies as a collaboration among men working from within existing academic disciplines but attempting to widen the boundaries imposed by conventional methods of inquiry.[6]

By the time Smith's essay appeared, vast changes had occurred throughout the world, political, ideological, intellectual. A tiny object called Sputnik had sent its "beep beep beep" around the world, and driven American complacency out of the skies. A new group of historians, including Daniel Boorstin, David Potter, Louis Hartz, and Richard Hofstadter, argued that consensus, not conflict, characterized American history. Still a younger school, labeled the "new left," had reacted strongly against the consensus.[7] By now they could not hope to find any concerted help from the American Studies Association. It had, without expressly working for it, become part of the establishment.

Most of the men who in the 1940s had declared holy war against "established" departments had quietly slipped into departmental niches themselves, and advised their younger colleagues to do the same. It is one thing to observe, with Howard Mumford Jones, that the departmental system (itself less than a century old) splits us into little groups conducting internecine wars, and quite another thing to restore the peace.[8] To put it differently, American Studies, often unable to acquire departmental status itself, dependent on the sufferance of the very departments from which it had to recruit students and faculty, has usually been quietly allowed to die of malnutrition.

From the first, members of English departments had led the movement. (Are historians more conservative and guild conscious than many of their colleagues? The matter bears further investigation.) Some of the most effective pioneer work dealt with reevaluating writers such as Melville, Poe, James, Twain, and Faulkner. Professor Robert Spiller, who more than any other man kept the national office alive at the University of Pennsylvania, was a literary scholar; so were the editor of *American Quarterly* (Hennig Cohen) and the executive secretary of the American Studies Association (Robert Lucid).

But the primary feature of the new cultural pattern of the 1960s as Susan Sontag pointed out, is that its model product is not the literary work, above all, the novel. "A new non-literary culture exists today," she writes, "of whose very existence, not to mention significance, most literary intellectuals are entirely unaware."[9] The basic texts for this new cultural alignment are found in the writings of Ludwig Wittgenstein, Marshall McLuhan, John Cage, Claude Levi-Strauss, Siegfried Gidieon, Norman O. Brown, and Gyorgy Kepes. They stress a new, more open way of looking at the world and at things in the world. Do they find a place in most English or American Studies courses? Hardly.

Of course it is not the aim of academic enterprises, old or new, to follow every new fad, but it is important to know how young people are viewing the world, what material they place a high premium on, how the gap between the past which scholars study and the present in which they function can be bridged. "Woe to the teacher who isn't competent and relevant!" warns John M. Culkin, interdisciplinary

Director of the Center for Communications at Fordham University in New York. "The students want the big thing, the real, the now. The generation gap may be the gap between intellectual and emotional development. What gets starved in the traditional culture comes out in the popular culture."[10]

Such statements would seem pivotal for scholars seriously concerned with American culture, but I doubt if there are a dozen programs in the United States, or abroad, which explore Culkin's "generation gap" in more than a cursory fashion.

One formidable problem lies not inside the academy but the culture itself. As Gertrude Stein reminded us, America is the oldest nation, being the first to enter the twentieth century. That meant it was the first to confront the new technology, alienation, and automation; the first to explore the terra nulla of the circuitry. This gives American area studies a different dimension from others and makes the old nineteenth-century thought patterns—still used extensively in graduate research and dissertations—hopelessly inadequate. The difference is the environment. It took man centuries to move from the natural to the mechanical, then, almost instantly, to electrical. Paradoxically, with change too fast to be discernible, everything is suddenly visible: segregation, poverty, senators' expense accounts, pork barrels. In an important recent collection of student essays (edited by Otto Butz) called *To Make a Difference,* Janet S. Schaefer, in her essay "Beyond Categories," argues that we must be willing quickly to relinquish categories and stereotypes if we are to understand America. Passenger pigeons, take note.

The job of every new generation is to seek new modes of expression. When the time span of "generation" is cut from twenty-five years to ten, which the new century has done, it is difficult to tell a mode from a fad, and impossible to age gracefully. So great is the acceleration that one is "out" before one fully realizes that one is "in."

Modes may be verbal, aural, visual, physical—an expression, a sound, a look, a gesture. It is the never-ending battle of Ancients versus Moderns, set to ragtime (except ragtime has been "out"). The Ancients voice the time-honored complaints: barbaric, fraudulent, erotic, inhuman, satanic. The Moderns have a much harder chore. They must coin or invent new modes, so that the hot may be cool and the innuendo the crescendo.

Whose law (Gresham's, Parkinson's, Darwin's, McLuhan's) applies? Who can stand on this cutting edge of culture and interpret what he sees or senses? Who will heed the insistent minority plea for comparative cultural research that would move American Studies from chauvinism to comparison, and develop a whole new set of landmarks by which the territorial imperative of a truly relevant field could be determined?[11]

Darwin speaks to the whole academic community, not merely the biologists. To survive, a movement must be able to adapt to new environments, to struggle with new problems. Many courses being taught, and theses being written, under the rubric of American Studies are attempting to answer questions no one is asking. They are demonstrating the same kind of routine, pedestrian coverage which made the founding fathers take action in the 1940s. In Darwinian terms, such courses do indeed invite comparison to the sad fate of the passenger pigeon.

Today's world is changing at a rate which defies description, let alone comprehension. The world alters as we walk on it. I have returned from a lecture tour in Eastern Europe, the fourth in four years. The young people there are talking, reading, dressing, dancing differently from those I met before. Not only my memories, but photographs, texts, and letters prove it. We *do* live in a changing world and are wired for sound. Neither bamboo nor iron curtains avail in the age of circuitry. All men, to some degree, are bombarded by the new media. The school, church, and universities are no longer inside buildings, but wherever sound and sight can penetrate. Recitations and resurrections have different dimensions. Phoenix-watchers, take note.

We begin with a single question, and end up with a number of additional ones:

> What are the conditions for the survival and growth of American Studies, here and abroad?
> If the movement does survive (as probably it will) can it also prevail?
> If the medium is the message, what does that message say to interdepartmental scholars?
> Why does American Studies remain strongest under the eastern establishment?

Why are our offerings well received in some foreign countries (Japan, Italy, Norway) and strongly resisted in others (India, France, Denmark)?

Why do other area studies (Far East, Islamic, Latin America) make headway in our universities, even as American Studies falters?

How much cohesion is necessary to survive in academe?

How do we instigate and then perpetuate the skeptical relentless furious search for truth, whatever labels we carry?

To ask such questions is both to doubt and to wonder. The sense of wonder, like doubt, is an angel which will let us repeat what Isaac Newton said of his own achievements: the great ocean of truth still lies all undiscovered before us.

The Popular Culture Boom

For half a century Americans have sung and whistled the theme song from an all-time favorite popular musical *Carousel:* "June Is Bustin' Out All Over." As the century ended we could add a variation: "Popular Culture Is Booming." Whatever else is global, American popular culture is.

Not everyone, at home and abroad, is pleased. Local, regional, and national interests are threatened. The battle is raging on a number of fronts. The final outcome is uncertain, but the boom is a reality. Land anywhere, and you will hear it.

Popular culture is generally considered "now time." Actually it is fascinated with both the future and the past, at the same time. This is the popular culture paradox. Two back-to-back recent events illustrate this. We send a satellite to Mars and re-enact a Civil War battle. Quite a paradox!

One hundred sixty-four definitions of "culture" appear in the volume by A. L. Kroeber and Clyde Kluckhohn, *Culture: A Critical Review of Concepts and Definitions.* Definitions and re-definitions keep pouring out. One of the most amusing is J. P. Morgan's: "Culture applies to families that have more than two servants." Then there is the breezy term *pop culture,* heard often nowadays. In H. W. Fowler's *Dictionary of Modern Usage* (1965), the only definition of "pop" cites pop music offered by symphony orchestras. Since then the term pops up everywhere, in all kinds of contexts.

We are beginning to see that pop has a long, complex history. We need to know more about how it functioned in ancient Egypt, Greece, Imperial Rome, medieval Europe, colonial Asia, and Africa. Recent scholars have offered clues. W. M. S. Russell has published "Plutarch As a Folklorist" and Timothy E. Gregory, *Vox Populi: Popular Opinions and Violence in the Religious Controversies of the Fifth Century A.D.* Much more work needs to be done.

Popular culture is more than fun and games. At its root is power. We have developed a superpower resting on superpopular culture. A much wider (if sometimes grudging) consensus concurs that as a new century and millennium begins, America is the only superpower. Many Americans even believe that, as with Britannia before us, America now rules both the air and the waves. If we count aircraft carriers, nuclear submarines, bombs, and bombers, that claim cannot be contested. The Soviet Union has collapsed, and other possible contenders, such as China, are still only beginning to emerge.

We have the power and with it the perks: predominance in media, movies, clothing, music, even food. Do these and other American items mean that we also have a super–popular culture? Is this our real secret weapon as we move through the twenty-first century?

Those whose power and perks have faded think it might well be so. Britain's Lord Hugh Thomas, in his 1996 book *World History,* concludes that this American domination offers the prospect of a better world than any other that can reasonably be imagined. This takes into account that the world might be compelled to copy our style of government, music, fashion, and food. Some evidence supports his suggestion. While this book was being written, McDonald's opened its fast food restaurant in its one hundredth overseas nation.

The Greeks had no word for culture. The first modern use of the word *culture* I can find dates from 1510, and has to do with the Latin root *colere,* to cultivate, or break the soil. Western European writers long used "civilization," not "culture," to denote social cultivation or refinement. Many still do. Roman language, law, and engineering were widely exported throughout the ancient world. One word, *populus* (the people, the majority), would later be used in popular culture. With their numerous holidays and use of "bread and circuses," the Romans invented what we now call the entertainment industry. They gave us a great empire, of and by the people, from the earth they inhabit and cultivate.

We know little about the *populus,* the majority, in the premodern era. Bits and snatches occur in letters or sermons, but usually "they" were referred to by their betters as "the mob." Thucydides and Tacitus assumed that common people were irresponsible and violent. In *The Republic,* Plato holds that democracy, the rule of the mob, is the worst possible form of government. Even in America, where centuries later Thomas Jefferson committed us to "certain inalienable rights," his powerful opponent, Alexander Hamilton, is said to have scoffed, "Your people, sir—your people is a great beast."

All that changed when anthropology, a relatively new study emerging in the nineteenth century, expanded and exploited the word culture. Today all sorts of "culture studies," such as American Studies, African Studies, Islamic Studies, Women's Studies, Lesbian Studies, Black Studies, dot university campuses. How do the ever-growing numbers of subcategories (elite, popular, folk) and subcultures (relating to ethnic, racial, geographical, and gender differences) function and affect such an inclusive term? Culture has become one of the most overinflated and overused words in the language.

Follow Emerson's advice, and "sit at the feet of the familiar, the low." We might also heed Emerson's sometime-gardener, Henry David Thoreau, who loved "the music of the telegraphy wires," and the insights of Walt Whitman, who shouted his barbaric yawp from the rooftops of the world.

East meets West in sunny Hawaii, long independent, now an American state. The East-West Center promotes multiculturalism and diversity, the natives do the hula-hula, the Americans dress up in tropical garb, and the Japanese run the economy.

The East-West Pop Link

East is east and west is west
And never the twain shall meet.

Rudyard Kipling
"The Ballad of East and West," 1889

Wrong, Mr. Kipling, but not entirely so. We have learned that the utopian dream of a Global Village which flourished in the late twentieth century, was a bridge too far, that racial, political, economic, and historical differences wouldn't end up in the blender and emerge unified. Dissent moved down like a wolf on the fold; in the twenty-first century, the global economy faltered and old-time nationalism and regionalism roared back, along with some unexpected problems: terrorism, potential chaos, and warfare in the Middle East.

But East and West did link up in one crucial area: popular culture. This important fact deserves our probing. Call it the pop link by which various trends, movement, and fields can be incorporated to bring the international scene into focus.

The nations of the world have long known one another by their elite productions. Now that the international style is upon us through the network of global mass media, we need other approaches, other materials that go beyond mass media. We need to understand that tradition works hand in hand with modernity in the rich mix of people, places, and perceptions.

How does one study such things? Where shall we begin? These are only points of entry, this scheme covering twenty topics or areas of popular culture:

1. National Integration and Popular Culture
2. Language and Ethnicity
3. History and Mythology
4. Tradition and Hierarchy
5. Theater, Outdoor Events, and Holidays

6. Film and Photography
7. Television, Video
8. Radio, Phonograph, Audio
9. Lifestyles—Food Habits, Clothes, etc.
10. Popular Literature (oral and written)
11. Sports and Recreation
12. Music, Jazz, Rock, etc.
13. Painting, Design, Sculpture
14. Architecture, Public and Private
15. Religion, Godmen and Godwomen, Pilgrimages
16. Advertising and Propaganda
17. Gender and Race
18. Cultural Geography and Geopolitics
19. Oral Folklore Tales
20. Rituals and Rites of Passage

The very nature of these topics requires firsthand observation, contacts, and experiences. We can never really penetrate the lives and experiences of other people by sitting in libraries or at computer terminals. Nothing takes the place of firsthand experience and observation. Peasant and primitive societies have very complex rituals and symbols. Much material that is first rate is found in the so-called third world.

We miss, or misinterpret a great deal on quick, hectic visits. In *Waterbuffalo Theology,* Kosuke Koyama says that before going to rural Thailand he thought that Buddhism was dying in the face of science and modernity. After three years in the field he completely reversed his opinion. He had actually experienced how Buddhism works.

Bearing all this in mind, Dr. Richard Gid Powers and I founded a new journal called *International Popular Culture.* It was born out of curiosity about the effect of modern communications on human cultures. The magazine has been created out of a sense of the perils as well as the promise of internationalism, a concern for the survival of national cultures besieged by international entertainment and lifestyles. The international exchange of ideas, entertainment, and goods has made it true (in a new and unexpected way) that nothing human is any longer completely foreign to anyone, anywhere. Can human cultures survive when the distinction between national and foreign is dissolved? What happens when nations import more and more of their culture from abroad? When the content of consciousness is foreign, will the patterns of consciousness, attitudes, and values also become foreign? When nothing is foreign, is anything native? Do humans everywhere become strangers in their own lands? Or does the internationalization of culture have precisely the opposite effect? Does the impact of international culture make a country more conscious of its own national identity, forcing a reexamination and reinvigorization of its own cultural roots?

Few fields hold such possibilities for international research and comparisons as does popular culture. Dramatic changes in media and technology have quickly expanded the field. High or elite culture speaks primarily to the well-educated minority. They were basically print oriented, so that the electronic revolution did not mean a radical change. But for popular or vernacular culture, the spectacular increase in communications efficiency had an equally spectacular substantive effect. The natural, organic, spontaneous part of vernacular culture continued to be local in orientation and became known primarily as folk culture. But artists and entrepreneurs who consciously created cultural products in order to meet a market demand found themselves faced with extraordinary opportunities. Railroads, telegraphs, automobiles, telephones, record players, radios, movies, television, and airplanes increasingly made it a relatively simple matter to market and sell mass entertainment on an international level. Thus, popular culture became distinct from folk culture as the former became an international phenomenon that transcended the nation-state.

No country benefited more from all of this than the United States. The power of popular culture penetrated where economic and military power failed. If Sparta was a military state, and Iran a religious state, the United States might be called a media state, beaming its message to the whole world. American music, first jazz, then rock and roll, and recently country; American television shows; youth dress, blue jeans, sweatshirts, and printed T-shirts; food and drink, hamburgers, cola, and wine coolers—all can be found in most parts of the world. Only a few isolated enclaves have resisted this onslaught. In the Soviet Union, teenagers offer tourists large sums of money to buy the jeans they are wearing. A Disney amusement park opened in Japan and is extremely successful. The Indian film industry makes hundreds of movies in Hindi with Indian actors and actresses who are placed in stories appropriate to America but totally removed from traditional Asia. Popular culture has developed far more of an international market than has high culture. Ballet and classical music were restricted primarily to Western civilization. They made few inroads in the third world. The popular culture which originated in the United States probably has had maximum effect on Western civilization, but, nevertheless, its effects on non-Western societies have been profound.

Thus, one might hope that scholars specializing in aspects of American popular culture would compare and contrast their conclusions with those of scholars examining similar phenomena in other societies.

Many questions need to be answered, problems need to be explored. Why do similar artifacts and activities have different meanings in different cultures? If American western movies are popular because we recently had vast open spaces and a frontier, why are they so popular in Italy and Japan? Why do Scandinavians relate so quickly to jazz? Arabs to Coca-Cola? Why have dozens of European and Asian countries fallen in love with American basketball but only a few with baseball? Asking and answering questions such as these will help students of popular culture separate the particular from the general and identify more precisely the relationships between surface behavior and underlying meaning.

Too often Asian history, like that of the rest of the world, has been written from the top down. We learn the names of the dynasties, emperors, pagodas, and temples but know nothing about the untold thousands who labored and died creating them.

A primary goal of popular culture is to include every epoch, people, and class and to be free of racial, gender, and political bias. Everyone has a place in the story of humankind. Popular culture is an aspect of democracy.

In American culture there is an increasing realization that Western culture without its strong Asian and Middle Eastern elements would be unrecognizable. The travels of Marco Polo to the Far East and the European discovery of Chinese and Middle Eastern arts and techniques opened eyes to the Greco-Roman heritage by stirring interest in other cultures.

We are beginning to understand how much we can learn from the Pacific Community.[1] In China, Korea, and Japan, Zen philosophy provides a clue to the difference between a unified culture vision and the battle of elite versus popular we experience in the West. This battle rages in the secular world as matters of good versus bad taste, and within our religions, as the sacred versus profane, the battle of good and evil. Zen offers the idea of the sacredness of all, the simple stone equal with the splendid jewel. Artistic care is given to the "ordinary" matters. The popular is also the profound.

In this way the East's cultural view is predisposed toward popular culture while the West has centuries of dualistic thinking to overcome in order to embrace everything people do as being worthy of attention, of interest, of appreciation; to accept everything human. What a discovery to find that such a culture exists and can bring new insight to the West.[2]

Few countries seem as ready to provide new insights as South Korea. The last of the "hermetic kingdoms," her doors were not opened to the world until later in the nineteenth century. The year 1982 marked the centennial of diplomatic relations between South Korea and the United States, but ties have become closer with every passing year. The interests of four major powers—the United States, the Soviet Union, China, and Japan—converge on this ancient and beautiful peninsula. Meanwhile, a series of economic and cultural successes, culminated in the 1988 Summer Olympic Games in Seoul, have given South Korea an ever larger role in world affairs.

Many unique features in Korean culture might be explored: the martial arts of tae kwon do or Korean wrestling *(ssirum)*. Special holidays include Tano, a day of feasting and events such as wrestling, and Ch'usok, the harvest moon festival. Hangul Day, October 9, commemorates the establishment of the Korean alphabet by King Sejong in 1446.

Because of the importance of cartoons and comics, we should examine that aspect of Korean popular culture. Kim Hong-do, a famous eighteenth-century artist, is said to have been the originator of the Korean cartoon; the first group of professional cartoonists emerged in the mid-1920s. At that time, "Bong-i" and "Kim-byol-jank-i" were carried in the monthly magazine *Yadam* as the first completely plotted cartoons. They were followed shortly after by the first newspaper comic strips, "Mongtongkuri" ("The Fool"), by No Shimsun in the daily *Chosun Ilbo,* and the works of Woongcho in the monthlies *Shindonga* and *Jokwang.*

World War II provided the impetus for comic book development in Korea. Kim Yong-whan, whose character is "Kojubu" ("Mr. Nosey"), took the lead in drawing completely composed cartoons, followed by Kim Song-whan ("Old Man Gobaoo"), Shin Dong-hun ("Notol Jusa" or "Mr Guffaw"), Ahn Eui-sup ("Dookobi" or "Mr. Toad"), Chung Woon-kyong ("Aunt Walsoon"), Lee Sang-ho ("Galbissi" or "Mr. Skinny"), and Kil Chang-dok ("Jadong-i" or "Clever Boy"). Some of the strips and books continued throughout the Korean War, the longest-lived of the strips being "Old Man Gobaoo."

Comic art is very popular in South Korea. In fact, the boom in magazines with about 800 titles has been a godsend for cartoonists. The number of comic strip fans increased considerably during the 1970s, and more than one-half of all newspaper readers turn to the comic strips first. Many of these strips satirize problems of a nation going through rapid industrialization, i.e., issues of pollution, inflation, or terrorism. Although children's comic books are relatively old in South Korea, it was not until 1977 that adult versions appeared. That market has continued to expand through the 1980s, and today Korean cartoons and comics are among the best in the world. They deserve careful scrutiny and analysis.

What sort of cooperation and collaboration might we seek? There are many possible answers. One good example turned up on my recent lecture tour of Egypt, the Adham Center for Television Journalism in Cairo.

The Center's goal is to introduce into Egypt international standards for television news field reporting, writing, and videotape editing. The Center teaches students how to use the electronic camera, how then to write a news story as "script" for pictures, interviews with natural sound on tape they have secured, how to read their story as a narration "voiced-over" picture, and how to edit that picture to the narrative sound track. Courses are taught in Broadcast Journalism, Television News Production, Introductory and Advanced Television News Reporting, and Writing. These courses are offered along parallel lines as a senior-level undergraduate concentration for Mass Communication majors, as well as on the graduate level. Perhaps this format, with suitable variations, can be set up throughout the Middle

East, Africa, and Asia. Then a vast source of information and understanding should result.

We watch; we wait. Nothing is more foolish than assuming we know what the future will bring. The stunning drama in China during the summer of 1989 showed how quickly events change the course of a nation, perhaps the world. Words of American patriot Patrick Henry, spoken in Richmond in 1775, echoed through the streets of Beijing—"liberty or death."[3]

What we can see is how much has happened, in the United States and the world, since Frederick Jackson Turner put forth his nineteenth-century frontier thesis, "The significance of the Frontier in American History." In the twenty-first century, we may well ask William Butler Yeats's question in "The Second Coming": "And what rough beast, its hour come round at last, / Slouches towards Bethlehem to be born?"

The Great Tradition

The fair sum of six thousand years' traditions of civility.

Coventry Patmore
Principles in Art, 1889

For centuries, an all-encompassing monomyth, built on a Judeo-Christian foundation honoring Greco-Roman roots, held Western cultures together. It's a sort of mythic-celestial glue. What might popular culture have been like then? In ages dominated by tradition, what might a traditional man or woman have believed, and what might they have considered popular?

Let's call this epoch the Great Tradition[1] and, for convenience, watch a single participant, whom we shall call Traditional Man, representing not males but humanity.

For more centuries than we can number or document, Traditional Man believed in mysteries he could not fathom. The repetition of archetypal gestures dramatized this; their truth he never questioned. His whole culture was "popular." He loved carnivals with their clowns, charlatans, and montebanks. Fairy tales were a constant and permanent delight. The Ugly Duckling, Jack and the Beanstalk, Snow White, and local saints were as real to him as his parents or neighbors. They were timeless because they participated in a transcendent reality.

Our "Now" is so vivid that we have not seen how much of our new wine comes in old bottles.

One is struck, scholars point out, with the tiny scale of life in early times, especially with the small size of the group in which nearly everybody spent their entire lives. There were few hotels, hostels, apartments for single persons, fewer hospitals, and almost no young men and women living on their own. In this landscape of meadows and open fields, with village communities scattered among them, there were many groups but only one basic mythic interpretation of life. And (if one defines "class" as a number of people banded together in the exercise of collective power) only one class existed, only one body of persons capable of concerted action over the whole area of society. The heads of the poorest family were the heads of *something*. They were not in a *mass* situation. England was a large rural hinterland, attached to a vast metropolis through a network of insignificant local centers.

A bit of that world was still left in my English grandfather, William Cross, who told me about his own youth, games, and holidays. A mechanic who loved his tools, he measured achievement by the strength and skill of his hands. He moved from the little town of Wincobank in Yorkshire to the New World in 1914, but he brought the Old World with him. In so doing, he became one of my major links to popular culture. He is a crucial tie not only to my past, but to that of all of us in cyberculture.

We can attempt to explore, in that preindustrial world, births, marriages, and deaths; personal discipline and social structure; social structure and revolution; the pattern of authority with its vast complexities. What was it like, one muses, to live an entirely oral life? In a world dominated by reading and writing, what was it like to face crises without the institutional patterns and supports we have come to take for granted? With few records of opinions, it is difficult to be certain how they viewed their own lives. Scholars of popular culture today confront the task of building an adequate foundation on premodern material and assumptions.

Most of them think of mass culture as the product of recent mass production and distribution. Some conceive of it more narrowly yet, as media entertainment. Recently a few have begun to challenge this approach, and to explore popular culture before printing. They insist that a whole body of important material has been ignored or misinterpreted.

A leader of this new school, Fred E. H. Schroeder, tells how he came to challenge the accepted parameters. In a Chicago museum, Schroeder examined small figurines that had been made centuries before in the Near East and China. He saw that there were other "Great Traditions" in other parts of the world that paralleled, but differed, from ours. The common link was popular culture and community.

Different traditions, different roots. Ours include the classical heritage, medievalism, the Renaissance, Scientific Revolution, and the Enlightenment. Other important items sustained the Little Tradition:

folklore, devotional images, mystery places and superstition, farces, festivals, and broadsides. Well into "modern times," the sixteenth and seventeenth centuries, both traditions shared a common source. It was the soil out of which tradition, meaning, and stability grew. The two traditions were interdependent. Great writers, such as Shakespeare and Milton, could draw freely from both.

Traditional Man drew inspiration from nature, not media; he worked with other humans much more than with machines. He was rooted and regulated by the weather and the seasons. He knew what he knew.

Life is made vastly more difficult by the Electronic Revolution—the so-called second or White Industrial Revolution, which depends not on coal but electricity. Enter the Global Village, which eradicates much that used to be local regional, or national. No need to go to Italy for a pizza or Hamburg for a hamburger; check the yellow pages of your phone book. In an age dedicated to multiculturalism, references to class, sex, language, or racial difference are taboo. We face the same attitudes in New York or New Delhi, in Prague or Peoria.

Sights and sounds that "turn people on" cross over all the old barriers. The Far East is next door, even in our garage. Rudyard Kipling was wrong: the twain have met. Radios, records, films, and satellites have world audiences and are replacing traditional newsbearers and storytellers. The Berlin Wall and many others have come tumbling down.

Good or bad, laudable or lamentable, we are experiencing a totally new kind of tradition in the making. Just what does "popular" mean in this context? What universal elements do they share? How do they reflect feelings and affect relationships?

To find answers humanists and social scientists must work together. It's the old struggle between emotion on one hand and logic on the other, a dilemma which was already ancient in Plato's day. We are still trying to bridge the gap.

Who will arbitrate, regulate, modify our Brave New World? The rules being suggested and debated are essentially unenforceable. Anything and everything goes. Privacy has disappeared with our old typewriters and old machines. We have sacrificed privacy for speed. Must the government step in?

In *Areopagitica* (1644) the English poet John Milton wrote his classic attack on government licensing of the press and on censorship. That became the accepted doctrine of free speech and the free press, but these doctrines don't fit the electronic media. The media over which messages and money flow are fluid and uncontrollable. Who then will be in control? Will the printed press be left free, but not the electronic media? What will happen to the freedom democracies have so prized for centuries?

Surely we don't want to halt the flow of information, any more than we want to deny access to water or electricity. We can no more block the incoming technology than King Alfred could block the incoming ocean tide. But we must look beyond the surface glitter and novelty. The new questions are not merely technical, but also social and ethical. What is at stake is human freedom and growth.

When old mosaics or mythologies disintegrate, new ones originate. To understand, we must be like the god Janus (for whom January is named) and look both ways at once. To comprehend the new, we must know where it came from and why, and then participate in it. Ready or not, we confront postmodernism. It has created new environments of impersonal and invisible power—new patterns, new problems, new space, new hype and hope.

We find ourselves in a free fall, borrowing, bending, and mending, debating about leaders, allies, methods, and the "canon." The new technology has upended the old cosmos, moving at the speed of light from reality to virtual or hyper reality. Our government has commissioned IBM to build an ultrasuper computer to do 3 trillion operations a second and retain 2.5 trillion bytes of memory. We are bringing time and space into a single entity. The foundation of traditional authority is being deconstructed and reconstructed. Where will this take us? Ready or not, this will be our Brave New World. Will it prosper or blow up?

Massive glut will be inevitable, and clever advertisers will manipulate and exploit. Audiences will be fragmented, quality will be diluted, and technology will be trivialized. Children will be especially vulnerable, as the marketplace becomes their message even before they can read or write. Holed up with a home information center, cut off from community and friends, individuals will be lost in cyberspace, victims of a new pervasive unreality.

Of course we cannot be sure about these things. Revolutionary change is always frightening, especially for those raised in an earlier (print) culture. Machines can't control us if we control machines. We program computers. They need never program us. We don't have to be slaves to 200 new channels, half of which ask us to shop and consume. We can turn off the set.

Theories of energy, matter, space, and time are in flux. What was once thought to be simple has become incredibly complex. Our future depends on our understanding the past, and those who have created and guided our democracy. Ralph Waldo Emerson told us over a century ago what type of people we need to fulfill the American dream: "Men and women of original perception and original action, who can open their eyes to new truths . . . people of elastic and moral mind, who can live in the moment and still take a step forward."

Let us take that step forward thoughtfully, trying not to fall on our face.

LOOKING AROUND

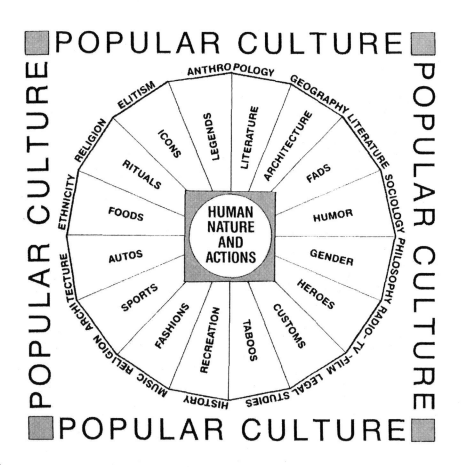

Pop Hype

Hype: Slang, implying excess or exaggeration; or a narcotics addict.

Webster's New Collegiate Dictionary

We wake up every morning, rushing in hurry-up time, and we take a bath in hype. How can we avoid it? It saturates our conversation, writing, television, chat rooms, AOLs, e-mail, newspapers, bumper stickers, and (to use a favorite hype-term) "much more." How many ads and urges do we see a day? Punsters act as if they know (a typical estimate is 1,500). Accurate? You tell me.

Hype by any name smells just as sweet—humbug, con game, rip-off, hoax, fraud, swindle—but the variant I like best was featured in the July 1995 issue of high-hyping *Playboy:* "Ballyhoo Boo-Boo and Boondoggling." Top that one.

As hype grows, so does the fascination with it. The word has been so overused, turned into jargon and clichés, that it is apt to turn up most anywhere and anytime. I just checked out my computer to find out how many "hype" entries are listed there: the answer is 808, and growing.

Not that we are the first to explore the field. In 1841, Charles Mackay published *Extraordinary Popular Delusions and the Madness of Crowds.* He centered on the South Sea Bubble that devastated Europe. Just think what he could do with the high-tech stock mania of the 1990s, which stripped billions of dollars from the American scheme and left hundreds of thousands jobless—dot.com to dot.gone. Or the endless stream of hyped-up patent medicines, guaranteed to cure whatever ails you. I have a late-nineteenth-century poster advertising "Warner's Safe Nervine" and listing the potion's wonderful qualities: "Gives Rest and Sleep. Cures Headache and Neuralgia, Vertigo or Dizziness. Positive Remedy for Nervous Prostration Caused by Excessive Pains, Drinking, Mental Shocks, Overwork, Etc., Etc."

A century later we get warning after warning from federal agencies about drug miracles (such as diet pills) that are new-day hype. We shouldn't be surprised. Recent studies show that hipsters, hucksters, and self-promoters have flourished for centuries.

America had its own master of the field: Phineas T. Barnum (1810-1891) who called himself "The Prince of Humbugs" and wrote books on just how he deceived and exploited a gullible public. Barnum's American Museum in New York City featured, among other things, Tom Thumb, "The Midget Marvel," Jenny Lind, "Sweet Songbird," and the nation's first live hippopotamus.

In 1871 Barnum bought a circus and turned it into "The Greatest Show on Earth." I still recall when, as a child, I watched his gaudy multicolored train roll into our Roanoke station to unload snarling lions and tigers, prancing clowns, trumpeting elephants, then parade them down Main Street, as the steam calliope belted out marching music. Yes, this *was* "The Greatest Show on Earth."

Has the greatest show moved to Las Vegas, where the annual Comdex-Fall trade show is the new vast electronic circus of product hype? Kim S. Nash describes it in her book *The Hype Masters.* All sorts of new technologies are on display, with carnival-style barkers (as in the old side shows) "educating" reporters on why their technology is best. There is even cross-hype between computers and the fashion industry, with slinky models and slick promises. There are "exclusive" parties; plenty of alcohol and a giant television screen where hyped-up pitchmen with teeth as large as diving boards shout out the New Gospel.

Why all the hype? Venture capitalists want in on big kills. Reporters want to get the big story. Punsters go to get the new puns. There's always something being sold as the magic bullet. And another super convention awaits, just down the road.

But the true permanent home of the hype is television. Fueled by flashy alluring ads, pictures of SUVs perched on mountain tops, happy happy women dressed for cocktails while mopping the kitchen, and kitties begging for one special brand of cat food, we become shell-shocked. Game shows show greediness at its worst, and intelligence at its nadir. Any who can get Oprah's stamp of approval become instant best sellers. Not only the ad gurus do all this but groups such as the competing politicians, the White House, the Office of Technical Assessment, astrologers, and (I am sad to say) The National Committee for Excellence in Education. No wonder the enduring nickname is "boob tube."

At special events, such as the Super Bowl (played in a superdome and featuring superstars), thirty-second ads can cost over a million dollars. Why? Because millions are watching, seeing magic moments on instant replay time and again, consuming gargantuan amounts of the beer and other items in the ads.

Have hype, ballyhoo, and disinformation driven the truth underground? Is the computer (as Joseph Wiezenbaum of MIT believes) "a solution in search of problems"? Under the surface are major propositions to ponder:

> Are we as mindless as the message we endure?
> Is the reckless violation of privacy the loss of freedom?
> Is the tone and content of electoral politics the loss of democracy and a competent governance?
> Who will tell the truth and admit profit decline?
> Has hype become an epidemic without a solution?

Hype has a first cousin which is also worth probing: *hyper,* the omnipresent prefix applied whenever we mean excessive or exaggerated. We deal in hyperbole. And hyperbolize. As if tension isn't bad enough, we have hypertension. If the poet adds an extra syllable to the final foot or dipody, creating a hypermeter, that means he or she is hypercatalectic. To say so is hypercritical. Let's hope this doesn't lead to hypertosis or hyperesthesia. Am I being hypersensitive? Might I be hypertonic? Enough!

Thomas Jefferson thought that if we gave the people light, they would find their way. A more skeptical person might say they would find their hype.

Then there is always Hollywood, Hype Central. Peek shows evolved into movies and the whole world wanted to peek. We always like to peek. That's what makes us human. So strike up the band, turn on the floodlights, bring on the performers. Hype, hype hurray!

The age of Barnum faded away, and Hollywood took over, with its supermovies, giant pageants, and visual extravaganzas. This suburb of Los Angeles got a new nickname—Tinsel Town—and a series of movie moguls replaced the earlier humbuggers. They even invented a new name for their medium, the magic screen, and the "star system," which remains the cornerstone of the new millennium's popular culture. The faces may be new but the formulas remain.

Then a powerful new force appeared on the scene: television. You could stay at home and get all the hype you could stand for free. Marshall McLuhan called television an extension of our central nervous system. Watching the army of couch potatoes crunch for hours, we think he might have been right.

We all know how hype seduces us. Computers are going to make living and learning easy. Buy two or three! Smart roads, run by computers, will eliminate gridlock and speeding; they are still trying to complete one outside of Blacksburg, Virginia. It would take us safely and quickly where we wanted to go. The folks up front would have the radio and cell phones; the kids in the back, television. Welcome to asphalt and information superhighways!

We built the super-interstate highways, and they proved to be death traps. Gridlock got worse and many new problems arose. Should the asphalt highway be a cautionary tale for the information highway? Are they both tales about the unexpected consequences of technology?

Raw data are clogging our arteries, and highways are taking us not to libraries but shopping malls. There you will find a vast wasteland of electronic junk, topped by games such as Mortal Kombat. Technology doesn't give us what we want and need, but what we're expected to want. Are you happy on the new Information Highway? Are you keeping an eye out for the breakdown lane? I am.

The Gap

In 1508 the great Renaissance artist Michelangelo was called to Rome by the pope to paint the ceiling of the Sistine Chapel. It was a staggering task for a man who was primarily a sculptor, and it took years to complete. The result: one of the most famous art shrines in the Western world, and images (recently cleaned) that seem to grow ever more powerful and relevant as the centuries pass.

Michelangelo set out as part of his task to retell the story of Genesis in the Bible: the creation of man and woman, their fall and expulsion from the Garden of Eden, and their agony as they faced the wilderness of the world with its continuation of sin. God reaches out, in this famous depiction, to touch man, whom he wants to "create in his own imge." Both hands reach out but do not touch. There is a *gap* between their two forefingers, and thereby hangs the tale.

That gap has been used innumerable times to express such things as heroic effort, false expectations, and tragic failure to be what God had first intended us to be. Time and again, in our own time, we have the image of God reaching out to touch (and sell) books, computers, pencils, television sets, furniture, not only to endorse them but somehow to grant them divine favor. That this might well seem to earlier ages a form of blasphemy doesn't bother the ad makers. After all, it works.

Sometimes we have two people—say, politicians, athletes, or candidates for election—"reaching out" again, seeking in some strange way to understand and trust each other. Another "gap" advertisement students always puzzle over shows a young woman standing on a toilet, painting Michelangelo's "gap" on her bathroom wall. Who sponsors the ad? Drum Chewing Tobacco. What is the connection?

The Man-God gap catches our imagination when applied to any number of things: gender gap, racial gap, income gap, language gap, age gap. At the moment, it highlights the tragic splits in Congress, and our national agenda, that have caused great anger and bitterness. President Lyndon Johnson urged us to "Come sit and reason together." We seem instead to sit and squabble.

We talk peace but wars rage. Over a dozen nasty struggles continue year after year, in Europe, Asia, Africa. We condemn and threaten but do little or nothing to close the gap. Genocide has become an expected (if not accepted) practice. In Cambodia, Pol Pot is said to have killed more people than Adolf Hitler. There, and in Kosovo and several African countries, pictures of the helpless victims document the crimes.

The religious gap worldwide seems to get ever wider and bloodier. Jerusalem is a tinder box ready to explode in the strife-ridden Middle East. Hindu and Muslim mobs kill and loot in India. Christian and Muslim gangs engage in bloody fighting in Indonesia. Many victims are stabbed, beaten, or trapped in fires. Others are attacked with rocks, machetes, clubs, even bows and arrows.

In Japan, long the model of order and decorum, the nation's schools are exploding with unrest; 44 percent of elementary and junior high schools report "collapsed classrooms" according to a report issued by the Kyodo News Agency. Once-obedient students fight, walk on desks, spit, urinate on verandahs, and talk nonstop during classes. What does this breakdown of civility mean?

Such gaps in world society do not seem to concern many Americans. After all, our CEO salaries soar, our car sales boom, our rich get richer. We suffer from affluenza: having so much of everything, being so affluent, facing endless glut and credit card debt.

Perhaps, more than everything else, it is wealth that divides the world and widens the gap. The United Nations' Human Development Report for 2002 makes this crystal clear. Last year the world consumed more than $24 trillion in goods and services—six times more than the figure for 2001. And consumption continues to skyrocket.

But this wealth is not spread equally around the world, or in our own nation. The United States, with less than 5 percent of the world's population, controls a staggering 40 percent of the world's wealth. The three richest people in the world own assets which exceed the combined gross domestic products of the world's forty-eight poorest countries. Hard to believe but true.

Of the world's 6.8 billion people, over 4.4 billion live today in what we call euphemistically "developing countries." What many are developing is poverty, AIDS, genocide, and despair. The figures on these unfortunate people numbering in the billions are both sobering and staggering. They should be contrasted to the "Happy Days Are

Here" rhetoric of our politicians—typified by President Clinton's State of the Union Speech on January 19, 1999. It's easier to get applause than results if you promise everybody something.

Three-fifths of the world's hapless 4.4 billion have no access to basic sanitation, and one-third have no safe drinking water. The world's children suffer the most. One-fifth of them are undernourished and cannot hope to get as far as grade five in school. How will *they* fare in the world of high tech, World Wide Web, and compulsory computers?

Ah, you say—along with those sharing our baby boom bonanza, many of whom make not thousands but millions of dollars a year— we can't afford worldwide basic education. It would cost $6 billion a year. We might remind them that in the United States alone we spend $8 billion annually on cosmetics. Add these to the list: Europe alone spends $11 billion annually on ice cream while in Europe and the United States we are spending $12 billion a year on perfumes.

What about the hungry? The myriads that are starving to death with no relief in sight? Again, the plump and prosperous (one estimate has one-third of Americans overweight) might point out that basic health care and nutrition would cost a staggering $13 billion. Does it bother them to know that Europe and the United States spend $17 billion a year on pet food?

Even harder to sanction are the billions spent on items which are known to have dire effects on users: $450 billion on cigarettes in Europe (the figure in the United States is not available); $105 billion on alcoholic drinks in Europe; and $400 billion on narcotic drugs worldwide. How do these expenditures affect the gap?

In times of astonishing prosperity and growth in America the Utopian Dream takes over. Few recall the word *utopia* comes from two Greek words—*ou* and *topos*—which mean "no place." In the end, Newton was right. All that goes up comes down. The Three Little Pigs also found out that to build our houses poorly and spend our time singing doesn't work. After World War II, we spent billions rebuilding Europe. We need such generous thinking and planning again. Instead, the world is full of war talk, finding a ready home in popular culture and flag mania. Popular support for military solutions abounds. We speak of "Evil Empires," pointing at North Korea and Iraq; they retaliate with similar phrases that stir their people.

Behind these threats are terrifying nuclear weapons that could kill not thousands but hundreds of thousands of innocent people. This in turn sets the United States, still hurting from the sneak attack on September 11, 2001, planning a missile intercept system that will cost billions. The Russians warn it might even reactivate the Cold War. Now there's a gap for you!

We know now how close we came to a nuclear showdown in 1963 with the Cuban missile crisis. At the last moment, our adversaries blinked and the missiles were removed from Cuba. We may not be so lucky with new enemies and threats.

Finding a way to close some of these gaps is not only a way to make a better world. It might be the only way to preserve the one we have now.

To Hack or Not to Hack

For many years the word *hack* had several clear meanings, such as a person unskilled at a particular activity, such as a tennis hack; a short, dry cough, or hack. It might be short for hackney—a carriage for hire with a hack driver. The word was extended to include literary drudges and political blowhards, or indeed anyone performing only for commercial success. (Does this make best-seller authors hack writers?) In the Age of Information (and disinformation) we are beset by this kind of hacker. The ghost of Hamlet's father pops up: words, words, words. Recall the 2000 presidential election.

Hack writers pollute the language, the way coal-burning smoke-stacks pollute the atmosphere. They combine elaborate jibberish with bits of applied psychology, utopian jargon, and trendy turns. Some of their slogans catch on and "stick." We get hooked on hack. Stale, flat, and unprofitable hack writers abound. Their number is legion and grows like weeds in a garden. "Hack" seems to be institutionalized with those writing "official reports," academic articles, legal briefs, and political speeches.

New definitions include an expert at programming and computer problems, or one who illegally gains access to and sometimes tampers with information in a computer system. All this raises ethical questions, and therein hangs our tale.

Pop culture attracts hack writers. So, for that matter, does elite culture. Scholarship has been turned into a hacker's paradise. Some of the worst comes from "professional" communicators, the kind Richard Hoggart describes in *On Culture and Communication:* "Communicators combine attitudes of late-Behaviorism with a touch of the Messianic fixer, laced with elaborate jargon compounded of some applied psychology plus some neurology plus some technology." Too often they end up like T. S. Eliot's hapless victim: "That is not what I meant at all."

Have we institutionalized hack writing in higher education, our degree factories? Publish or perish: so what do we publish? Dull, trite articles and books fit only for a few peers and promotion committees.

Thus scores of bright original thinkers have become hacks. Once they are addicted they find it hard to kick the habit. Being hooked on hack is hobbling some of our brightest young minds, spreading like a plague through higher education. Why don't the addicts complain, rebel, defy? Some do and get fired. Others leave the profession, to go where good writing is prized for its own worth.

Academe has its own falling stars, with titles such as "public intellectuals" and "the overclass." Russell Jacoby coined the phrase "public intellectuals" in 1987 and *Time* followed suit with a splashy feature on "The 100 Top Overlords and How They Induce Us to Do Their Bidding." Academic celebrities! Soon people such as Cornel West, Camille Paglia, and Doris Kearns Goodwin were all over the conference circuit, featured on television and in such magazines as *The New Republic,* the *Village Voice,* and *Dissent.* How did these "public intellectuals" move from trendy gurus to relics in one short decade? Who popped their balloons? We need to probe.

Stanley Fish blew the whistle in his 1995 book *Professional Correctness,* calling the new stars "cameo intellectuals" who were as shallow as the mass media talking heads. Then Richard Posner published *Public Intellectuals: A Story of Decline* (July 7, 2002). The tide was turning. Jean Bethke Elshtain summed it up neatly: the public intellectuals were getting more and more public and less and less intellectual. The cycle of popularity was working again, as it always does. Their fifteen minutes of fame was up.

New trendy creatures haunt the halls of academe; they have different subtitles but the generic name is Cult Studs (Culture Studies). What holds most of them—American Studies, African Studies, Afro-American Studies, Diet Studies, Women Studies, Islamic Studies, Gay Studies, Environmental Studies, etc.—together? The titles are the message: they are often out-and-out advocacy groups, well meaning but often into hard sell: multimedia, on-camera posing, with a dash of golly-gee. They are proud pious pilgrims, marching resolutely into cyberspace and taking no prisoners.

They envision a world out of balance, out of sync, and believe that they alone get to set it straight. They seldom seek other points of view. They love to mount rallies and protests, full of mobile meanings, shifting connections, brief encounters, and what they like to call (in a mysterious phrase) intertextuality. They are desperate to represent and attract "the people." They never consider the fact that "the people" don't, in the main, understand just what they want or why. Never mind. Cult Studs go marching on.

Is there a hacker ethic? Can we explain and apply it? A 2001 book, *The Hacker Ethic,* by the young Finnish hacker, Pekka Himanen, provides an interesting answer. Hackers are heroes who are opening up windows for a new world. They move us away from the Protestant ethic (as described by Max Weber) built on money, work, stability, and accountability. The new ethic is based on passion, enablement, freedom, and openness. Hackers are idealistic pioneers laying the cornerstone for our new economy. Think of them as explorers, inventors, guerillas, and joyous adventurers, changing the world and the way it works.

Displacing the Protestant work ethic will take time, as with all great cultural changes. In a historical context, Himanen believes his ethic resembles the pre-Protestant ethic more than the Protestant one. No longer will work be an end in itself: similar to our pre-Protestant ancestors, hackers have fun, leisure, and festivals. His hacker heroes are Bill Gates, Bill Joy, Ken Thompson, Dennis Ritchie, Peter Samson, and Steve Wozniak—saints in the new electronic religion. His book ends by adapting a line which has been used by young revolutionaries for centuries: "Computer Power to the People!"

Not all "the people" have been happy with this new computer power. James Rule (writing in *Dissent*) calls the computer frenzy "the business equivalent of Viagra: an alluring new snake oil, pretending to alleviate obstacles to productivity; a quick fix." Instead of emancipation, we get depersonalization. Bill Joy, one of the pioneer hackers, has also expressed deep doubts about the revolution he helped launch.

Donna Britt, *Washington Post* columnist, is another skeptic. One day, she writes, her new computer froze. Similar to millions of others, her job now depends on computers. Neither she nor a baffled technician could revive hers. In desperation, she called her highly technocratic brother, Bruce. Surely he would solve the problem. Bruce was distraught. That same day a hacker had usurped his password, read all his personal mail, got his e-mail addresses, and entered charges on his compromised credit card. So much for the pioneering hacker ethic. Frustrated and helpless, Bruce was more desperate than Donna. He was left with the same feeling he had had when someone robbed his apartment, a feeling of violation.

Both Britts recalled even more serious break-ins. Only a month earlier a group of hackers called the "Keebler Elves" broke into the National Severe Storms Laboratory computers, vandalizing a Web site officials use to check daily storm activity nationwide. Among the thirteen Web sites tampered with by the Elves were several run by NASA, the Army, and the U.S. Department of Education. The Feds were baffled.

The year 2001 saw a number of startling events (all involving the new electronic technology) which severely damaged our national security, and thus our future. Suddenly computers and data were missing from Los Alamos National Laboratory, where our most important nuclear secrets and weapons are supposedly impenetrable. The true consequences have never been revealed.

Security in Washington, DC, fared no better. We learned to our dismay that Robert Hanssen, a mole in the most secret sections of the CIA and FBI, had used his computer skills to spy for the Russians since 1985. U.S. intelligence officials are still evaluating the damage caused by this superhacker, who handed over some 6,000 pages of documents to Russian intelligence. A presumed highly successful "secret tunnel" under the Russian embassy in Washington also had been compromised by Hanssen. Not only had we spent millions of dollars to no avail, but the Russians had turned the tables, fed us false information via the tunnel, and caused untold damage around the world.

One of the books salvaged from Hanssen's house was David Major's *U.S. Counterintelligence, Ethics, and Conflict.* A lifetime friend of Hanssen, a stunned Major could only say, "Obviously, he didn't read my book." Or did he simply read it, scoff, and reverse Major's hacker ethics?

The contrasts between Pekka Himanen and Robert Hanssen's "hacker ethics" are so great that they must give us pause. Himanen believes that hacker culture is nothing less than a fundamental breakthrough in the discovery of the world unfolding "in the certain dawn of the third millennium."

9/11

Seduced by our high-tech years of fast economic growth, and a seductive new "-ism"—globalism—we rushed into a new millennium and new century with fanfare and frenzy. What happened? History did what it always does. It happened.

When?: on a balmy, cloudless, Tuesday morning, September 11, 2001, in the land of the free and the home of the brave. Popular culture shortened it to 9/11. What happened that morning, we have said ever since, would be the most documented event in our entire history. Our world was turned upside down. Our dreams became nightmares. New York, the jewel in our crown, saw its World Trade Center collapse as steel, concrete, blood, and guts covered the streets of lower Manhattan.

Washington was similarly upended, as we watched portions of the Pentagon burst into flames and turn to ashes when a second hijacked airliner dove into its walls; heard reports of yet another hijacked plane, possibly headed for the Capitol, downed by American martyrs in rural Pennsylvania; saw Congressmen and Congresswomen spontaneously join hands on the spared Capitol steps and sing "God Bless America."

Globalism made way for a new "-ism," terrorism, not only in America, but worldwide. It struck when and where we least expected it. We were traumatized, terrified. We still are.

Words once used by a few scholars and specialists—Taliban, bin Laden, al-Jazeera, al-Qaida, Kandahar, suicide bombers—were in every paper, on every channel and newscast. On a single bright morning, 9/11, had World War III begun?

Fear fought freedom, and fear was winning. "Security" was the new buzzword. Bears ravaged Wall Street. Mutual funds that were the new peaks became the valleys, with lifetime losses of many billions. Psychiatrists were deluged with frantic patients.

There were positive aspects. Millions of Americans responded. Some stood in endless lines to donate blood; others sent supplies by the truckload for rescue workers. Countless people devoted time, prayers, energy, and money. Celebrities organized a telethon with donations going to victims' families. New York's mayor Rudy Guiliani led many New York services in a splendid display of action and work hours beyond the call of duty. Post 9/11 may well have been *our* "finest hour."

While many nations also reacted generously, problems arose. Relations between nations and former allies became more complex, fragile, even hostile. Global harmony became global tug-of-war. We had sadly underestimated the power of tribalism, nationalism, love of language, fear of change. History and tradition show how grossly naive it is to think we were in a new utopia. The highly touted Global Village started to look like Global Pillage.

Who would save the ailing airlines, monitor the airports, settle gigantic insurance claims, rebuild a shattered New York and public confidence? How could and would we win a war in Central Asia? Make a new modern nation in Afghanistan? "You will be sitting ducks," recently defeated Russian generals warned.

Flags flew in abundance, churches filled and overflowed, old "Guts and Glory" movie reruns appeared on television. People were warned they must be patient. Teenagers, never patient, developed a 9/11 slang; black humor that might deflect the horrors and ongoing fear that engulfed them. A messy room became "ground zero," a handsome boy "firefighter cute," out-of-style clothes "a burka," a fellow student being dismissed from school "was a total jehad," and if someone was caught up in something petty, "That's so September 10." The edgy pop-centered slang tried to turn dark tragedy into comic relief.

But who could say what might happen next, and when the people would start to sing a different song? These were indeed new times, and they try our souls. What can we say that has not already been said often time after time? Who dares to pass a final verdict on what 9/11 meant and will mean for years to come?

What we can do is treasure and ponder some of the remarkable images resulting from 9/11 at what came to be known as Ground Zero. Ours were made by George Marr and have never been seen before. They are powerful. Let them speak to you, as they have to me.

This cross of iron beams, which fell intact from Tower One, was discovered standing upright by a recovery worker two days after the tragedy. The twenty-foot-tall cross became a symbol of hope to the rescue workers toiling at Ground Zero (March 8, 2002).

Flags and banners from around the world bearing messages of sympathy and hope line the iron fence of St. Paul's Chapel, across the street from Ground Zero (March 8, 2002).

Blowing in the wind, scraps of paper from a WTC tower office dangle in a tree in the cemetery of St. Paul's Chapel, near the viewing platform for Ground Zero (March 8, 2002).

Ground Zero and ongoing recovery work as seen from the public viewing platform (March 8, 2002).

Tribute to ironworkers; sculpture in front of the Federal Building near Ground Zero (March 8, 2002).

Visitors solemnly examine tributes and messages of encouragement left in front of St. Paul's Chapel near Ground Zero, just before the six-month anniversary of the terrorist attack (March 8, 2002).

Thousands of tributes and memorials to the victims of September 11, 2001, line the fence and spill onto the street outside St. Paul's Chapel. Tributes were periodically removed and taken inside for safekeeping (March 8, 2002).

Ground Zero panorama (July 13, 2002)

Cyberlore

Once an America mired in Depression asked, "Brother, can you spare a dime?" Now, over seventy years later, the cry is "Brother, can you share a silicon chip?" In the same California where the 1849 gold rush began, another one of far greater proportions is taking place in Silicon Valley. Now the "workers" go not into streams or caves, but to new buildings filled with electronic wizardry to "pan for gold" or, in their term, "push for high-tech." Leading the pack is Bill Gates, our new J. P. Morgan, the multibillionaire who presides over Microsoft where most of the original employees (so the story goes) are now millionaires.

There are millions of instant millionaires whose new fortune is based on silicon and not gold. The new wealth has revolutionized the economy of many other nations around the globe—especially Japan. From goldlore to cyberlore—this may be the most appropriate title for our new century. When did "cyber" get into our culture, and how is it changing our world?

No computers were available when World War II began. During that war, we developed one for military purposes to speed up ballistic calculations—a monster named Eniac weighing fifty tons.

"Cyber" comes from the Greek word *kybernetes,* meaning helmsman or guide. Cybernetics is best defined as the scientific study of those methods of control and communication that are common to living organisms and machines, especially the analysis of the operations of machines such as computers. The word has expanded, similar to the computers that are its trademark, into our offices, factories, homes, schools, economy, and mythology. To many, they are symbolic and iconic. To others, they are a threat, perhaps even diabolic. To everyone they are a consuming new element of the New Age.

As do many others, I struggle to place "cyber" into perspective, and to measure its impact. I am writing this chapter on my computer. If we need to "talk," it may well be by e-mail. Surely high tech will be involved in producing the book, and orders for it will come in electronically. Like it or not, I am adrift in the cyber sea.

We are all too close to it, too involved with it, to make definitive and permanent judgments. My general attitude is skeptical. That will have little to do with how others fail or flourish in cyberspace. As always, history gives us the one wise answer: wait and see.

While we are waiting, we look for clues, signs of what awaits in cyberspace. The many blessings are well touted and advertised. Magazines, books, and myriad sound bites urge us forward, faster and faster. Some of my friends now claim (rather proudly) that they can no longer write except on their laptops, that they welcome oncoming digital signatures, since they can barely sign their name the old way. A pity.

Of the many skeptics and critics, none has had more impact than the late Neil Postman. Two of his books are *Amusing Ourselves to Death* (1985) and *Technopoly* (1993). Both have become rallying cries for intellectuals and theorists. Here is lore to bring us headlong into the twenty-first century. Will we end up with a new dominant lore—technolore? If so, we had best first examine technopoly, on which it will rest.

* * *

Just what is technopoly? The first explicit principles were laid out in Frederick W. Taylor's *The Principles of Scientific Management* (1911). The primary goal of human labor and thought is efficiency; technical calculation is superior to human judgment; and the world should be guided by experts. These principles, designed for the industrial workplace, eventually seeped into all aspects of American culture, and the "corporate lifestyle." It turns workers into "hands," then robots, becoming a kind of totalitarian technology, and its chief support comes from computers.

Computers don't work; they direct work. They have little value without something to control. Bureaucrats use computers to create the illusion that decisions are not under their control. "Computers have determined . . ." Has anyone in or out of the university not met this mandate? A bureaucrat armed with a computer is the unacknowledged legislator of our age.

Postman traces the rise of technopoly in his new and provocative book with that title. He builds on the work of Lewis Mumford, Herbert Read, and Jacques Ellul. All cultures, he contends, fall into three types: tool-using, technocracy, and technopoly. Humans devised tools over the centuries to meet immediate and specific survival needs: the arrow, club, spear, and even more complex items such as temples and cathedrals. Humans also invented fairy tales.

Fairy tales live deep in the human psyche, primitive and primordial. They are among our first and best teachers. We learn from them to fear wolves (one ate Little Red Riding Hood's grandmother), hungry giants (one chased Jack up the beanstalk), and wicked stepmothers (poor Cinderella). Consider for a moment the sorcerer's apprentice. He used his master's magic to get broomsticks to do his job of carrying water. It worked, but he couldn't stop the brooms. A flood resulted and disaster threatened. The angry master returned just in time to save the day.

We don't have sorcerer's apprentices these days, at least not by that name, but try this possible substitute: *technopoly*. The new popular buzzword restates the ancient axiom of Socrates—that the unexamined life is not worth living. Technopoly suggests that the unexamined technology is not worth having.

This fear of rampant, unchecked technology goes back to the Luddites, nineteenth-century English textile workers named after Ned Ludd, who smashed spinning jennies to protect the handicraft and cottage industries. The past century saw it surface in books such as Aldous Huxley's *Brave New World* (1932) and George Orwell's *1984* (1949). More recent writers, such as Lewis Mumford, Christopher Lasch, and Neil Postman, have sent out warning signals, and in *The True and Only Heaven: Progress and Its Critics* (1991), Christopher Lasch implies some truth may be found in John Randolph's definition of progress as a "luscious lie."

No one can deny that technology has done incredible things for us; in this era of silicon chips, space walks, AWACS, medical miracles, and virtual reality, the whole world is wired. The question is, what has all this done *to* us? How much of the highly touted Information Revolution is built on ad hypes, media distortion, television ratings, and patriotic chauvinism?

Technopoly has leveled its biggest guns at the computer, as it does many things (processing, sorting, and storing information) extremely well. Still, other things (thinking, feeling, creating ideas) it can't do at all. Mankind's noblest ideas (justice, equality, compassion) are not computable. Yet many people, including those in positions of knowledge and power, do insist on humanizing and empowering the computer to "lead the way." To them computers have "brains" possessing "intelligence." MIT's John Pfeiffer wrote a book calling the computer *The Thinking Machine* (1962), and John Kemeny has forecast a "symbiotic evolution" of the human and computer species. Many claim that computers are not only smarter but more durable than humans, that they might become the dominant species of "life" on Earth.

Think of the implications. Will *Homo sapiens,* whom nature evolved in two million years, and whom Moses, Jesus, Buddha, and Mohammed gave their lives to redeem, be doomed to oblivion by IBM and MacIntosh?

Part of the problem is linguistic. Computer people have developed a language of their own, unlike any of the human languages. With shades of Orwell's doublethink, once you learn one "system," the model or indeed the whole system changes. We live in the electronic age of enforced obsolescence. Whatever you have today won't be good enough tomorrow. You can count on it.

Meanwhile, pray that your computer doesn't break down, reject disks, erase months of work, or get a virus. If such a calamity occurs, and we hear of them daily, then it's off to the professional repair service. In due time (and at considerable expense) it will come back with a message attached. One sits near my office now. It reads: Internally Terminated. System 7.0.1. tune-up. At Ease 1.0. At ease? Not me! I prepared to fight other battles in my youth, and I know what follows "At ease."—"Attention! Shoulder arms! March!"

The key operative word today is NEW—new century, new digital imaging, new teleworking, new frontier, new models, new mergers, new guard, new wealth, new age. Cyberspace is today's fairy godmother; but is hype mixed with the hope? Cell phones drop calls, PDAs lose data, PCs crash. Spam clogs our e-mail and glut crams our offices and our lives. What's that buzzing noise?

But as hope rises, so does hype. Are you ready to move on to biochips, gallium-nitrate semiconductors, automatic computing, subatomic quantum computers-on-a-chip, and nanowires? Don't you want to stay in the game?

To be or not to be—that is the question. Meanwhile, one best hold onto any rechargeable batteries. They may come in handy.

Heroes: From Halos to Handcuffs

Oh, how Americans have wanted heroes, wanted brave simple
fine leaders!

Sherwood Anderson
A Story Teller's Story

History is meaningless without leaders and heroes. "It is natural to
believe in greatness," Emerson wrote. "All mythology opens with
demigods." Heroes are mirrors of the times and inspirations for our
future.

In America we have all but forgotten many earlier heroes. We find
that those we have trusted and admired have feet of clay. The heroic
vision has given way to scandal mongering, media hype, and elevated
celebrities who pass across the heaven like shooting stars, then disappear into the eternal darkness. "I'll never forget old what's-his-
name." We already have.

Very few young people identify with the heroes of their parents, let
alone their grandparents or ancestors. Andy Warhol predicted we will
have a "fifteen-minute culture," in which everyone will be famous for
fifteen minutes. In 2003, a startling change has occurred: we have ex-
changed the heroic halo for handcuffs. Those mighty tycoons who
have posed as heroes are carted off to prison in handcuffs.

We have what the *Christian Science Monitor* calls, in a front-page
headline, "The Return of Anti-Tycoon Rhetoric." Signing the new
corporate reform bill in Washington, the president had a new name
for a "heroic" class many lionized only months ago: "No more easy
money for corporate criminals. Just hard time." From heroes to criminals in one swift journey. What's left?

Quick—who's your hero, male or female? Can't name one? You're
not alone. We are in the antihero era. When did it begin? Perhaps with
the 1963 assassination of our young Prince Charming, President John
F. Kennedy. Kennedy's assassination, we now see, was the most crucial event in the heroic history of our generation. Because of the times

and technology, Kennedy had a global popularity unlike that of any other president or of any man then alive. That he should be killed senselessly by an ex-marine, who was in turn killed on television before millions of viewers, formed an unbelievable historic episode.

The "Great Against" set in with the 1960s, featuring slogans such as "Burn, baby, burn" and "Hell no, I won't go!" Groups such as the Chickenshits carried yellow flags, played kazoos, and parodied everything heroic. They liked to drop to the floor and crawl away mumbling, "Grovel, grovel, grovel. Who are we to be heroes?"

In his best-selling book *Soul on Ice,* Eldredge Cleaver accused every president since Lincoln of "conniving politically and cynically with human rights," inspiring widespread civil disobedience and riots. In *Heroes and Anti-Heroes,* Harold Lubin noted that "heroes rise with a dazzling brilliancy, but are extinguished with all the finality of a shooting star." In *Hero/Anti-Hero,* Roger B. Rollin went farther: "Heroes and villains are becoming interchangeable. You can't tell one from the other without a program." Bill Butler gave the idea a savage twist, writing in *The Myth of the Hero,* "We demand heroes to be exemplars, then to be our dinner." We had reached a heroic vacuum.

Bad news such as Watergate, Vietnam, and President Clinton's impeachment only increased the depth of the vacuum. Our economic woes, following the explosion of the great economic bubble of the 1990s, were a final blow. Down we tumbled.

What happened to the heroic vision on which this nation was founded and flourished for two centuries? Were our heroes and heroines lost, strayed, or stolen?

We entered the time of the Great Against, not only against the Establishment, the Man, parents, Reds, but also against God (rumored to be dead), virginity, and civility.

"What has happened," a Japanese observer wrote, "is that substantial and structural problems have so shaken American society and politics that institutions have lost their ability to restore themselves." In his essay "The Decline of Greatness," Arthur Schlesinger Jr. writes, "Ours is an age without heroes. We have no giants who play roles which one can imagine no one else playing in their stead."

First print, then radio, then telvision, then computers wrought incredible changes in our culture. They have taken us away from reality to virtual reality. Celebrities, heroic substitutes, often exist largely on celluloid or in cathode tubes. Polls and listings of leading celebrities flood the papers and media. I am looking at one published in August 2002. At the top is Britney Spears (wriggling body parts), Tiger Woods (hits golf balls), U2 (yells loudly), and Michael Jackson (moon walks). Not a hero or heroine on the list.

Throughout the ages heroes have been our guiding stars—trustworthy, truthful, solid. They never take the easy road but follow the truth wherever it may lead. Our enemies have their own heroes, and we must demand some of our own, both men and women who will bring light to our darkness.

Heroes used to look up to heaven; now they use it as their stage. Astronauts have not only created a new hero type, but a new area of concern: astropolitics. One of the original space heroes of the 1960s, John Glenn, described by *Life* magazine as "moral and strongly religious," was pictured before his flight shaping hamburgers for the outdoor grill; twenty years later, he was shaping his candidacy for the White House.

Heroes, as with everything else in our society, are getting more complex, more multifaceted. In our "Never go out alone" and "Exact change after dark" culture, it is not rapture but rape that characterizes urban living. And how can anyone control nuclear arms, restore the economy, clean up the environment, personalize the machine? We go to the movies to laugh at bumblers such as the late Peter Sellers or Woody Allen because we empathize with their plight. We like them because of, rather than in spite of, their incompetency. The same can be said for a line of television favorites, stretching from Wally Cox, Ozzie Nelson, Bob Denver, and Dick Van Dyke to John Ritter. Playing "The Greatest American Hero" in a television series, the "hero" tries to fly like Superman—but runs into a billboard.

What about Wonder Woman? Is she something of a clone? Or does she have qualities which I don't recognize? This could be antifeminist, this idea that a woman should act like a man. Perhaps that's what feminism is telling us.

In any case, we should take a close look at the comics if we want to understand today's hero and superhero. Why did Spider-Man (introduced by his creator, Stan Lee, in 1962) become the leading comic celebrity a decade later? Unlike the Hulk, Superman, Captain Marvel, and Wonder Woman, Spider-Man was a superhero in the liberal tradition, fighting drug pushers, organized crime, and pollution. By

1980 both the country and Spider-Man had grown a bit weary of crusades and crusaders. Does this attitude help explain the popularity of Ronald Reagan and his laissez-faire policies? What can we learn about other cultures by studying their comics?

Certainly comics have become a sizeable factor in the new frenzied world, and much more study of them is in order. I think we will find that Americans are the most Promethean; those of no other nation seem to depend so much on super torsos and super muscles.

Our rise to power from struggling colonies to world leadership in less than three centuries has strong Promethean overtones. The unconditional surrender of the Germans and Japanese at the end of World War II, and the sole control of the atomic bomb, made the twentieth century the American century. Have we lost our way in the brave new world?

Connecting the Dots

A brand-new culture cliché has burst upon the scene: CONNEC-
TING THE DOTS. The FBI and CIA failed to connect—did this make
the September 11 tragedy possible? Our economic gurus and CEOs
don't do it. Hence an up-and-down stock market and scandalous corporate management announced day after day. Our colleges and universities fail to connect dots and thousands of their graduates chase
an ever-shrinking job market. The Roman Catholic Church didn't do
it and faces overwhelming criticism and defection. Can anyone connect multiple Homeland Security dots?

Just what are these dots, and what is this widespread epidemic?
Was the great poet W. B. Yeats right—"Things fall apart, the center
cannot hold"? On a basic level, the "dots" are raw information, often
in the form of statistics, sound bytes, clever slogans, garbled dis-
patches coming in too fast to be digested, connected, or interpreted.
What we do not have is the pattern, which turns information into
knowledge.

We live with mass culture, which sells the sizzle rather than the
steak. It thrives on tinsel wrappers, false claims, shadows, more evasion, and phony fellowship. The danger is that we peddle cotton
candy when the people crave real substantial food. Compare "folk"
with "mass." Folk is active; mass is passive. Folk is the arrow, and it
sings. Mass is the target, and it listens. Folk shows and tells. Mass
waits to be told and seduced.

Awash in flashy slogans and information overkill, we cannot connect the dots. Neither can those who try to lead us. Walter Lippman
recognized this as early as 1922. In *Public Opinion,* he identified
"oversimplified patterns and allures in modern life which defend our
prejudices." They teach us to ignore the world outside and concen-
trate on the picture in our heads. Surface overwhelms substance. We
recall the mother who, having been congratulated on the fine baby in
the carriage, replies, "Oh, that's nothing. You ought to see her photographs!"

In such a world, how can we hope to connect the dots? That is the big question of our time.

Vox populi puts what Everyman likes on the TV screen, radio network, movie marquee, videotape, and billboard, in the kitchen, bedroom, classroom, and supermarket. Each thing is a cultural cipher. To decipher is to find the key to the culture in which people actually live. Images are not things. They are flimsy and weightless; they flicker by and we cannot catch them. They fade away like the morning dew. They are no-*thing*.

Semiology, the study of signs and symbols, takes linguistics as the model for studying culture artifacts—films, television, print, fashion, food, art, architecture. The basic question is: How is meaning generated, transmitted, and received? This is achieved only fractionally by words—which some either can't or won't acknowledge. We are "speaking" through our posture, gestures, facial expressions, eye contacts, clothes, hair, and an infinite number of other devices, even when we aren't "saying" anything . . . and what we *say* may be quite different from what we *mean*.

What about the secret structures in our minds, which affect the way we interpret signs and symbols? Might "culture" be defined as a codified system? Is it through codes that our thought is filtered, our value systems established, our attitudes about the world formed? Is this the place to begin a study of popular culture? Or should we begin with the simple *things* that best define us and our culture?

James Gleick, whose book is titled *Faster: The Acceleration of Just About Everything* (1999), is on the outer reaches of the electronic frontier, where he is both encouraged and frightened. He has a new religion. This is his credo, in his own words: "I may not believe in God, but I believe in the network as global village and global brain." We are not sure just what he does believe in or what new God he worships. Clearly Gleick tries hard to make Microsoft safe for capitalism. He goes for catchy chapter titles: "Open Sesame," "This Is Sex?," "Big Brother Is Us," and "Please Hold for Me, Ma Bell," for example. His style is snappy and choppy. This is part of the new "electrolingo," and Gleick has mastered it. Is standard English becoming archaic? Just what does this mean?

The information future seems out of control. Billions of dollars are invested in search of the future, but it's the present we have to cope with. Click OK to agree. Get online or get out of circulation. Meander forever in a chat room. Get on eBay and scrutinize the offerings. Bid on one of the 4,645 Furbys for sale today, or a car, or a house.

* * *

Behold the New Age, new century, new high tech! Bright enough to blind us like the midday sun. But the sun sets, and a dark side comes with the night. Although most choose the sunlight, some are examining the dark side. We have 9/11, snipers, terrorists, and things that go bump in the night. We all know what the new age promises, and the high tech promotes. What has it delivered? What happens when the sun goes down? How can we handle the new darkness? Here is a question ripe for probing.

Every book about the past is a product of the present. I am interested in the study of small worlds and individual people, but can we leave out our national and unified past when we look back? Must we not see the forest, not just certain single trees? This is the historian's dilemma.

It's a brand-new century. Many of us are still at the stage of unscrewing our solitary light bulbs whenever we plug in the fan. We are forced into the new electronic ritual, the never-ending exhausting dance with the invader. Technology traps us and we resist. We want to get back to the simple life, but of course, we can't. We envy the once-free Tarzan who swung effortlessly through the primeval jungle. We just live and flourish in our baffling over-dotted jungle.

Can we connect the dots? What a challenge!

Spoiled by Success?

> There is not so impudent a thing in nature as an assured man, confident of success.

> William Congreve

The great American success story of our time has been McDonald's—how Ray Kroc teamed up with Ronald McDonald to create the greatest food network in the world, flourishing in 121 countries in 2002 and hoisting the Golden Arches over culture after culture.[1]

And such humble beginnings! In 1946 two Canadians, the McDonald brothers, opened their first fast-food drive-in in San Bernardino, California.[2] Aging Ray Kroc bought out the restaurant in 1954, adopted a promising success motto (KISS—Keep It Simple, Stupid), and dotted the global landscape with new McDonald's.[3]

Success breeds success—the idea kept growing. But times change, winners falter, and hubris sets in. In probing popular culture, we must remember the glory days. For years, people went under the Golden Arches as if there were a gold rush. In some ways, there certainly was.

The beginnings were simple enough. Endlessly repeated is the rags-to-riches story that could well have been written earlier by Horatio Alger. Ray was a middle-aged salesman, peddling milkshake mixers in the 1950s. He got a big order from California, where the McDonald brothers, Dick and Mac, had opened a drive-in restaurant in 1937. They cooked hot dogs (not hamburgers), mixed shakes, and hired three carhops to carry food to parked cars. Soon booming, they moved, expanded, and hired twenty carhops in 1945; added fifteen-cent hamburgers to the menu; and emphasized speed and smiles. Kroc entered

I am grateful to Brad Trask of McDonald's International for supplying information and material, and to Christopher Rooney for help with illustrations. Some of this material was used in a different form in my book *Popular Culture: Cavespace to Cyberspace* (Binghamton, NY: The Haworth Press, 1999).

the scene and all his lights turned on. He bought their franchises in 1955 and their name for $2.7 million in 1962. The McDonald myth was in motion.[4]

Kroc opened his place in Des Plaines, Illinois, in 1955, and made the best decision of a lifetime—to use the twenty-five-foot yellow sheet metal golden arches logo designed by Stanley Clark Meston. This has become one of the most recognized signs in America, attracting millions at home and abroad.[5] The world had been McDonaldized.

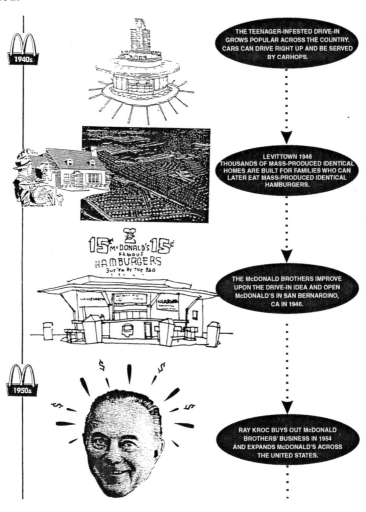

The 1955 prototype restaurant was modified to accommodate different climates and expanded service. Basements were added for furnaces and storage; the service area was enclosed in glass walls. The sheet-metal arches were replaced with backlighted plastic arches. During the 1960s the arches, originally designed as parabolas, were rounded off. The newly designed restaurants featured a low-profile mansard roof with brick and shingle texture. But the golden arches' popular commercial vernacular style had taken on a symbolic life of their own.

So had Ronald McDonald clowns as attention getters, the source of fun, laughter, and trickery.

> The official spokesman for the McDonald's Corporation, Ronald is the world's best-publicized clown. He dances, prances, and becomes one with children. He works magic, loves hamburgers, and defies the laws of gravity. Ronald McDonald Houses house families whose children are hospitalized. He is the very essence of goodwill.
>
> The first Ronald made his local debut in Washington, and his national debut a few months later in Macy's Thanksgiving Day Parade in New York City, November 25, 1963. He came on network in 1966, and was dubbed "official spokesman" in 1967. Over the years, by constant exposure, he has become part of our media landscape: the genial host of McDonaldland.[6]

The meteoric expansion and success of McDonald's has been described many times. I have told the story and written a book titled *The World of Ronald McDonald,* while George Ritzer described the *McDonaldization of Society,* in a world which features McDentists, McDoctors, McChild Centers, with *USA Today* our McPaper, serving us new McNuggets.[7] The golden arches had become as American as, well, the hamburger and our special "way of life" and new ideas of production and consumption. We live in a world of supermalls, superstores, shopping networks, cybermalls, Webs, online connections, and infomercials.

Now the tables are turned. Instead of attacking, McDonald's is being attacked, at home and abroad. On August 12, 1999, a Parisian-turned-sheep-farmer named Jose Bove led a French mob that ransacked and demolished a McDonald's restaurant in the town of Millau. French labor unions, Communists, ecologists, and national-

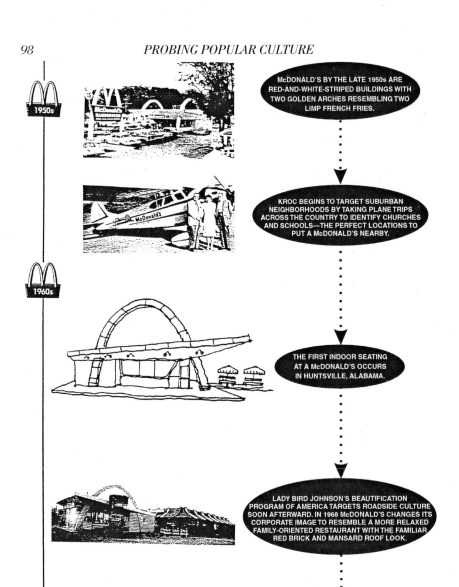

ists shouted their approval. The statue of Ronald was destroyed, along with the whole building. A kind of collective madness centered on the world's most popular food chain. Would McDonald's be a casualty of the new millennium?

The worst was yet to come. The attack on big Mac moved to our own shores in 2000 and centered in Norfolk, Virginia. There PETA (People for the Ethical Treatment of Animals) announced an assault on McDonald's that would include billboards, bumper stickers, posters, T-shirts, and print ads. They will feature a picture of Ronald holding a bloody butcher knife and a dead chicken. PETA's vegetarian co-ordinator, Bruce Friedrich, wants to talk about "the billions who suffer," not about McDonald's "billions and billions of hamburgers." One immediate solution: sell veggie burgers nationwide. The campaign began in Norfolk with the first billboards. There was a Web site, MeatStinks.com, and a two-page spread in the *Animal Times* magazine put out by PETA for its 600,000 members.

McDonald's spokesman Walt Riker called the PETA campaign unwarranted, adding that McDonald's remains "committed to animal welfare and will work with other groups on the issue."

McDonald's sales began to slip. Competitors with "more healthy foods" advanced. The time had come to change menus and images at a time of economic recession, 9/11, mass hysteria over terrorism and the snipers, and fear of an invasion and war in Iraq. There was little for Ronald to dance and prance about in 2002.

Then, in November, came somber news: McDonald's would eliminate 600 jobs and close restaurants. It would close restaurants in underperforming countries. The reductions would be achieved through attrition and the elimination of open positions. The New York stock market shuddered, and the Dow Jones fell on Wall Street. On December 17, 2002, McDonald's announced the first downward quarter in forty-one years.

Was the era of the Golden Boy under the Golden Arches over? What did this mean for popular culture, and the American Dream?

LOOKING BACK

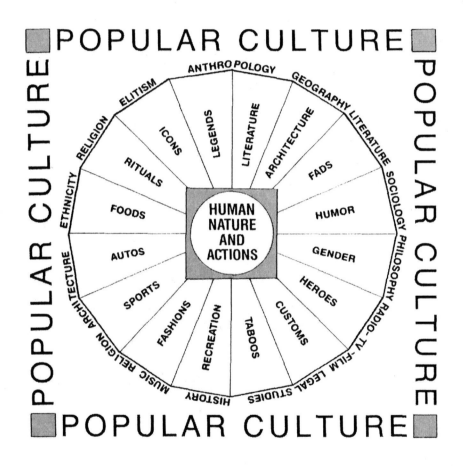

There he stands, six-shooter drawn, bringing law and order to a chaotic and lawless world. The American Cowboy is the enduring symbol and idol of the America he helped to patrol. What were the real cowboys like?

The Cowboy and World Mythology

I see by your outfit that you are a cowboy.

Popular Western Ballad

At first glance, it's hard to know who a person is or what he or she does, but everyone can recognize a cowboy or cowgirl. Their clothing, walk, and talk set them apart. They have a freedom, a code, an openness that we have lost. We want to get it back. Somewhere, deep inside, we want to be cowboys and cowgirls. So do people all over the world.

We read in Genesis about a Garden of Eden, free of cares and woes. Other cultures have their Shangri-la, their Nirvana, their secret hideaways. We once had our wide-open plains, seas of grass—the Golden West. There, one could be free, doing the "what comes naturally."

Jean-Jacques Rousseau captivated eighteenth-century Europe with his description of the natural man. That romantic symbol of unbridled freedom triumphantly entered the American forests as the buckskin-clad hunter, only to emerge on the Great Plains a century later as the American cowboy. Somewhere between the Alleghenies and the Rockies the followers of Daniel Boone traded coonskins for sombreros, long rifles for six-shooters, and moccasins for spurs, without losing for a second their fascination for the hero-loving American public.

The two symbolic figures, the hunter and the cowboy, made identical appeals to the trait traditionally valued above all others in the United States: freedom. The hunter wasn't happy unless he had what Daniel Boone called elbow room; which, when translated into the twentieth-century parlance, became a popular cowboy ditty titled "Don't Fence Me In." There was something nostalgic about the lyrics; for two decades the cowboy could roam for hundreds of miles with no fences to hamper him or his herd. Now those days are as distant as speculation over the morals of Grover Cleveland and the feasibility of eating tomatoes. With the introduction of barbed wire and thousands of homesteaders in the 1870s, the open range quickly disappeared in fact, but not in fancy. On the back of Old Paint, the cowboy has ridden through whole libraries of serious literature, hundreds of light novels, boxcar loads of pulp paper, and miles of celluloid (mostly grade B), alas, the popular hero par excellence of twentieth-century America. When John Glenn orbited the earth in 1962, the "cowboy's" new empty space extended for light years.

As the city traffic, drug traffic, pollution, and confusion grow, wide-open space becomes ever more alluring. Environmentalism becomes a major force, and tourism and rodeos grow by leaps and bounds. We have trouble remembering the names of our governors, mayors, and representatives in Washington. But who doesn't know that John Wayne is the Duke, and Clint Eastwood wants to make your day? No other occupation has a more loyal following than the cowboy does. Hollywood knows this full well. Every fourth movie ever made there has been a western.

What is their message? Why do we like to see them time and again? Out there, somewhere, justice triumphs. The wicked are cut down, and the good prevail. The physical setting? Will James tells us in *The Drifting Cowboy:* "There's still hundreds of miles of country where there's plenty of cattle and no fences, where the cowboy wears his boots out in the stirrup and not in irrigation ditches . . . where a man's a man."[1]

Famous New York newspaper editor Horace Greeley knew this when he gave his famous advice in 1859: "Go west, young man." It was Cowboy Country then, Marlboro Country later on, and still is John Wayne Country. Geography has given way to mythology, and a new breed of cowboys have ridden their spaceships into outer space. Range wars have given way to star wars. Underneath, it's the same story with the same characters. The Good Guys outwit the Bad Guys. We can all sleep soundly, knowing that justice has prevailed.

Long before John Glenn went into space in 1962, the cowboy had captured the world's imagination—not only with movies and novels, but also in the actual presence of World War II American GIs and sailors who circled the globe on the ground. Wherever they went they were expected to shoot the way the sergeant had taught them, but from the hip; to abandon their jeeps if a horse of any description was available; to ignore posture drills and let their legs assume a normal bowed position; and to toss army tactics aside to track down enemy bad men in posses. The popular presses and Hollywood movies had done their job well; the cowboy legend preceded the Americans around the world.[2]

This remarkable fact is even more noteworthy when one considers that the West's open ranges were gone forever before the turn of the century. Conditions have altered radically, but the stereotype of the open-range cowboy remained static. What is the essential history of this popular hero stereotype? Who created and perpetrated it? What function does the cowboy legend serve in our culture?

In analyzing the cowboy we are dealing not so much with specific individuals as with a recognizable type, not with a mere historical reality, but with a popular ideal. This was precisely the case, the Age of Enlightenment, with Rousseau's natural man; it was equally true of the man of the forest on the early American frontier. We feel that the West is governed by a compelling and unwritten code, a "spirit" unlike that east of the Mississippi—and that the cowboy constructed this code and epitomizes this spirit. Whether this is an oversimplification is beside the point. Many Americans think it is the case, which is basis enough for the stereotype. Behind this stock character is a core of cowboy history that is the foundation on which the legend has been constructed.

That history stresses this fact: to be close to the "real" thing is desirable; the cowboy is a good man to have with you, and a demon when he is against you.

In his own bully way Theodore Roosevelt was a publicist for the strenuous life. He fell hook, line, and sinker for the cowboy, as his book *Ranch Life and the Hunting Trail* (1888) shows. Describing the cowboy with the usual adjectives, Roosevelt calls him "the grim pioneer of our race," who prepares the way for civilization from before whose face he must disappear.[3]

Heroic feats of early cattlemen are plentiful. Oliver Loving (1812-1867), Ike Pryor (1852-1937), and Charles Goodnight (1836-1929) led lives that would make a Hollywood executive have Technicolor dreams. In 1866 Nelson Story (1838-1926), with a handful of cowboys, drove 3,000 cattle all the way from Fort Worth to Montana, right through the heart of Indian country, despite stiff opposition from the Sioux, Cheyenne, and United States Army. Larry McMurtry told the same story generations later in his Pulitzer Prize-winning novel, *Lonesome Dove* (1986).

The first man to utilize the cowboy in the dime novel had a single individual as a model: Prentiss Ingraham, Buffalo Bill's ghost writer, wrote about the cowboy Buck Taylor, who was a feature attraction in Cody's Wild West Show for years. In 1887 Ingraham ground out the first cowboy novel to appear in a popular series; it depicts a fictional Buck Taylor as a youngster living in a Texas Ranger camp, and eventually "winning his spurs" with the famous outfit.

No one could match Zane Grey (1875-1939) when it came to writing cowboy novels. He wrote over sixty. Grey had only one story to tell—that of the heroic cowboy surmounting all human and natural obstacles, to bring justice and peace to the community. America and the world wanted to hear the story, and he told it well. Many accepted it not as fiction but as truth.

Being "western" is, for thousands of writers, real estate men, restaurant owners, tourist specialists, and showmen, not only a characteristic but a vocation. It is hard to tell where genuine love of the cowboy stops and genuine concern with the pocketbook begins. Yet the greatest western publicist was a man who died debt-ridden and harassed by creditors. "Buffalo Bill" Cody, more than any other individual, turned an American region, the West, into a tradition. By living so dramatically and so successfully in the role he assumed, he convinced people that the West and the cowboys who set the tone and the pace were heroic. He is vastly more important in American history than most historians realize.

The single organized activity that has done most to glorify the American cowboy, and one fostered for that specific purpose, is the rodeo. A Spanish word, *rodeo* was a term used in the Southwest long before it was a nationally known term. No one knows when the first rodeo was held, but a number of them supplemented nineteenth-century fiestas in the Southwest. The event had traveled as far north as Cheyenne, Wyoming, by 1872, where the citizens were treated to an exhibition of Texas steer riding which the *Daily Leader* editor found "unnecessarily cruel, suited to the prairie."

The distinction of having the oldest continuous annual rodeo is claimed by Prescott, Arizona, where the event has been held since 1888. Promoters of the Cheyenne Frontier Days celebration (begun in 1897) advertise their event as "the daddy of 'em all," which has given Prescott no recourse except to build their publicity around the notion that theirs is the "granddaddy of 'em all." From these two sires have sprung progenies throughout the nation. Eventually the rodeo came east. One of the most spectacular found a home in New York's Madison Square Garden.

Cowboys' idols of Hollywood have been quick to cash in on the rodeo fad. Gene Autry, who so impressed his neighbors in the town of Berwyn, Oklahoma, that they changed the name to Gene Autry, Oklahoma, got together a rodeo group called the Flying A Ranch Stampede, which opened as the feature attraction of the 1942 Houston Fat Stock Show. Gene and his horse, Champion, were in the vanguard. The performers' costumes were treated with fluorescent dyes, so that they glowed when the concealed violet lights were turned on them. All in all, it was enough to make a veteran of the Chisholm Trail rush out to the closest swinging doors.

Roy Rogers, Republic Picture's singing cowboy, also developed a rodeo performance that was smooth and streamlined; but even in this sugarcoated and commercial form the appeal of the West is strong.

The meteoric rise of Hopalong Cassidy, in real life William Boyd, indicated that ether waves, television grids, and film reels were a most effective device for creating new American heroes. Born in Ohio in 1896, William Boyd left home in 1915 to seek his fortune in Hollywood. As a leading romantic star in the 1920s, he got a mansion in Beverly Hills, a beach house in Malibu, and (in a rather rapid sequence) four wives. Coming upon hard times during the Depression and the end of silent movies, he resorted to roles in class B westerns and Hopalong Cassidy scripts prepared by Clarence E. Mulford. By 1943 he had shot his way through fifty-four separate movies in the series, when the rise in production costs caused producer Harry Sherman to abandon the series. After dashing off another dozen movies on his own hook, Boyd conceded defeat and let the series lapse.

A few years later Boyd gambled all his remaining fortune on the television rights of the Hopalong pictures. He was able to acquire them and, when television expanded, to ascend to the role of cowboy hero par excellence for young America. Featured on sixty-three television stations, 152 radio stations, 155 newspaper comic pages, and the wares of over a hundred manufacturers of western specialties by the end of 1950, Hopalong had not only become a cowboy, but also a large-scale enterprise that netted a million dollars a year.

Not Hopalong but the Duke, John Wayne, is the Cowboy of the Century. Unlike earlier cowboy favorites, he got even larger in death. Over two decades after his 1979 death from lung cancer, "The Duke" remains one of America's favorite movie stars. The prototypical cowboy, he has become a culture icon with each passing year. Born in 1907 in Winterset, Iowa, he was a singing cowboy on low-budget films before starring in John Ford's 1939 classic, *Stagecoach.* One great Western followed another, made in Utah's Monument Valley: *Fort Apache, Red River, She Wore a Yellow Ribbon, The Searchers, The Man Who Shot Liberty Valence,* and *True Grit.* A man's ideal more than a woman's, Wayne earned his popularity without becoming a great actor or a sex symbol. In all his films, whatever the character, John Wayne portrayed John Wayne, a persona he created for himself: the tough, gritty loner whose mission was to uphold the frontier's and the nation's traditional values.

By a magic only possible in the ephemeral world of popular culture and cinematic fame, Wayne's private identity was replaced by the Duke—a man who did what a man has to do, representing America itself. Books about him since 1996 document his iconic stature, slow drawl, and "down-home" qualities. The books include Herb Fagen's *Duke, We're Glad We Knew You,* Ronald Davis' *Duke: The Life and Image of John Wayne,* Kinky Friedman's *God Bless John Wayne,* Garry Wills' *John Wayne's America: The Politics of Celebrity,* and Emanuel Levy, *John Wayne: Prophet of the American Way of Life.* Books and articles keep coming.[4]

As the open range closed, the movie cameras and TV opened to the cowboy and his myths. Hundreds of westerns poured out of Hollywood, full of singing cowboys, masked rescuers, gunslingers, rodeo stars, sheriffs, pristine young ladies (often schoolteachers from Boston), rustlers, and the bad guys in black hats. From the days when silent film cowboys had to depend on the viewer's imagination to hear their six-shooter pop—and when Bronco Billy and William S. Hart set the rules for the new genre—until the offbeat adventurers of midnight cowboys and unforgiven ones in the 1990s, the western was pure screen magic.

Blazing guns and thundering hoofs ruled prime-time television. The three major networks (ABC, CBS, and NBC) carried thirty western series in the 1950s, 1960s, and 1970s. The most successful, *Gunsmoke,* ran for twenty seasons (1955-1975), becoming the longest-running dramatic show in television history. The cowboy and popular culture have never had a better union and success. James Arness played the heroic Marshal Matt Dillon of Dodge City. Among Dillon's friends were Doc Adams (Milburn Stone) and Kitty Russell (Amanda Blake), owner of the Long Branch Saloon. Throughout its run, the series maintained a high standard of quality and excellence.

Four other series left deep marks on the media. *Wagon Train* lasted from 1957 to 1965. Ward Bond was the wagonmaster, Major Seth Adams, who was assisted by scout Flint McCullough (Robert Horton). Frank McGrath played cantankerous Charlie Wooster, the wagon train's cook and comic relief. Millions of fans tuned in week after week. *Maverick* (1957-1962) moved into the realm of humor and satire, but many episodes followed traditional Western paths. James Garner played Bret Maverick, a gambling ace who would do just about anything to avoid a fight but could handle himself physically if pushed to the edge. Bret shared the limelight with brother Bart, played by Jack Kelly. Bret and Bart usually alternated as leads, but at times appeared together. The show sometimes satirized other popular TV series, such as *Dragnet* and *Bonanza.*

Second only to *Gunsmoke* as the longest-running western in television history, *Bonanza* (1959-1973) was the first western to be televised in color. The show starred Lorne Greene as Ben Cartwright, the patriarch widower of three sons, each borne by a different wife. Set in Virginia City, Nevada, in the 1860s, the show depicted the story of a rancher family, owners of the huge and prosperous Ponderosa Ranch. Adam, the serious, intellectual oldest son, was portrayed by Pernell Roberts; gentle, naive "Hoss" was played by Dan Blocker; and the youngest son, romantic and impulsive Little Joe, featured Michael Landon.

Have Gun, Will Travel (1957-1963) depicted Richard Boone as a college-educated gunslinger whose calling card bore the picture of the white knight chess piece (paladin). The show was an immediate hit, and its theme song, "The Ballad of Paladin," sung by Johnny Western, was a hit single in the early 1960s.

The cowboy changes but his myth remains. As I write this there are almost 800 "cowboy books" in print and hundreds of earlier ones. Cowboy history and lore have become an industry, and "drugstore cowboys" roam the streets of towns that never saw a longhorn cow. Cowboy clothes (especially blue jeans) have a world market. "Country Western" tops the music charts, making millions for people who are neither country nor western. Merely browsing in recent cowboy books is fascinating. Among my favorites are Jon Scieszka's *The Good, the Bad, and the Goofy;* Guy Logsdon's *The Whorehouse Bells Were Ringing: And Other Songs Cowboys Sing;* John White's *Git Along, Little Dogies;* Katie Lee's *Ten Thousand Goddamn Cattle;* and Rick Steber's series, *Tales of the Wild West.*[5]

"If we could dispel the haze," writes Walter P. Webb in *The Great Plains* (1931/1981), "we could view western life as it was in reality—logical, perfectly in accord ultimately with the laws laid down by the inscrutable Plains." We can never dispel the romantic haze that has settled permanently on the western horizon, and wouldn't want to, even if we could. We like to conjure up our untarnished natural noblemen roaming about in their never-never land, where they make the laws and mete out the justice. We like to read about it, see it on the silver screen, describe it to our children. It is a world in which none of our problems occur. Everyone knows what he or she is supposed to do and does it. We have never really lost the tranquility and finality of the Medieval Synthesis as long as the tradition of the American West is kept alive. This is what the editor of the popular pulp magazine, *Ranch Romances,* meant when he wrote, "We aim to lead our readers away from the complexities of civilization into a world of simple feeling and direct emotion."

The American cowboy symbolizes a freedom, individuality, and closeness to nature which has become a mirage for most of us. When things get too bad, we turn on the TV, slip into the movie house, or into a chair with the latest cowboy magazine or novel, and vicariously hit the trail. We become free agents in space and time, and leave our humdrum world behind. "As I sat in the movie house it was evident that Bill Hart was being loved by all there," wrote Sherwood Anderson in *A Story Teller's Story.* "I also wanted to be loved; to be a little dreaded and feared, too, perhaps. 'Ah, there goes Sherwood Anderson! Treat him with respect. He is a bad man when he is aroused. But treat him kindly and he will be as gentle with you as any cooing dove.'"[6]

The cowboy legend is a tangible safety valve for mechanized and urbanized America lost in psychotherapy, teenage killings, disinformation, culture wars, and overcrowded cyberspace. We remember and cherish scores of western films: *Shane* (1953), *Yellow Sky* (1948),

Hondo (1953), and such offbeats as *Butch Cassidy and the Sundance Kid* (1969), *Blazing Saddles* (1974), and *Unforgiven* (1992). They reveal artistic conventions as rigid as those in a Greek drama or temple, a classical sonata, or a medieval morality play. Good and Evil finally clash head on. Everybody on and off the screen holds his or her breath. Guns blaze. Smoke clears. Destiny is suspended for a fleeting moment. The cowboy hero wins. He must. Defender of things that matter, he is unconquerable. He fights not for laws or gain or property but to prove that the universe makes sense.

The cowboy saga is the only American art form in which the ancient notion of honor retains its full strength. Since the cowboy is not a murderer but a man of virtue, always prepared for defeat, he retains his inner invulnerability. "His story need not end with his death, and usually does not," writes Robert Warshow. "What we finally respond to is not his victory but his defeat."[7]

If one had to choose two archetypal western films, they might feature the same cowboy, Gary Cooper. The first, *The Virginian,* was made in 1929. Taken from Owen Wister's novel, it featured the handsome and fearless cowboy who wooed and won a lovely innocent schoolteacher from Vermont, Mary Brian (Molly Stark Wood). The villain taunted and ridiculed him in vain. "When you call me that," the Virginian said quietly, "smile." All of us knew that sooner or later the evil Trampas (Walter Huston) would "ask for it." Gary Cooper would then see that he got it, for he had no choice. That was the whole point of this life and of the film. Honor, maidenhood, and justice had to be defended. They were.

Over a quarter century later, Gary Cooper starred in *High Noon* (1952), perhaps the ultimate western. He was still calm and virtuous, but older, less handsome and sprightly. Gray hairs mingled with black. The pattern, as well as Cooper himself, was maturing.

Urged by his Quaker wife, Amy (Grace Kelly), the well-respected marshal was prepared to turn in his badge, hang up his gun, and start a new life. Suddenly news comes that a vicious killer whom the marshal sent to prison has suddenly been pardoned. He plans to come to town on the noon train, and with three other outlaws, hunt down the marshal and kill him.

Cooper is forced to play out the drama and "do what he has to do." But unlike the Virginian, he knows he can never achieve an easy or absolute triumph. Justice is restored through his own still-superb skills and assistance from his wife, who shoots one of the outlaws. Then Cooper takes off his badge, throws it in the dust, and rides out of town with his wife. He stood by his oath, and brought law and order back to the West; but all he wanted to do, in performing the noble task, was to stay alive in the dusty street of a jerkwater cowtown. He had to live in both worlds at once. So do we all.

The cowboy, like the hunter before him and space traveler after him, holds up well in the new century. He has the virtues our culture most admires and seldom sees in political and financial leaders. With him the love of freedom and fair play is a passion, and his willingness to accept responsibility a dogma. To his cattle, his horse, his friends, and his code, he is always faithful.

The open range is gone now. Barbed wire won out. Only an imaginary trail winds its way through what was once a limitless sea of grass. But fond memories remain of that golden moment, and the names are still on the land: Bitterroot, Rawhide Creek, Stranglehold, Whoopup, Chugwater, Cheyenne, Dodge City, Tombstone, Horse Thief Creek, Purgatory Hill, and Medicine Bow.

Heartfelt and descriptive, they will not leave us. Nor will the memory, the hope, and the dream.

Paul Bunyan:
Fakelore Meets Folklore

Maybe the scholars have been following a false lead: Maybe popular literature isn't a folk art at all.

Bernard DeVoto

In the United States, land of the Big Build-Up, much that passes for folklore is really fakelore. Scholars have found that Paul Bunyan, Joe Magarac, Pecos Bill, and other "ancient characters" are (like Mercutio's wound) not so deep as a well, nor so wide as a church door. But what we have done with them (again like the wound) is enough; 'twill serve.

Only a purist, convinced that "the folk" are holier than specific contrivers, resents what has happened. A hero is a hero, no matter who creates him, or why. Too many folklorists think they are scientific when actually they are sentimental. They deplore efforts of corporations or political groups to invent new symbols and characters, without realizing that in our society these are the natural agencies to do such things. Like everything else, folklore and mythology are shaped by the culture in which they flourish.

A lumber company was the prime mover behind our modern Beowulf, Paul Bunyan. But that doesn't make him any less vigorous than old Beowulf, or Aeneas, or Samson.

Take that fellow Samson. Chances are, as one tough-minded folklorist recently pointed out, that he was nothing more than an overgrown Asia Minor country boy. He made his first unpremeditated bid for fame when he leaned against a shaky pole in the tent of some other desert-swelling character and in his awkwardness caused the shelter to collapse. Relatives and friends transformed the tent into a pillar and Samson's reputation was made.

Delilah? Probably just a pretty hillbilly who didn't get a wolf whistle from the Babylon bucks until neighbors started to spread tales about her. Then the boys looked her up, but they hardly thought students would be doing it in twentieth-century encyclopedias.

The whole thing makes one wonder if in a few generations folklorists will be rushing around getting old-timers to tell what they remember about Lydia E. Pinkham, Al Capone, Bing Crosby, Babe Ruth, and Mae West. By then, the Babe will have become as sturdy a figure as Samson, and Mae will have Delilah beaten on several points.

Fakelore and folklore are delightfully intertwined in Paul Bunyan.[1] *Fortune* magazine chose him, a few years ago, as "the one fictional character fit to stand with the nation's historical heroes." For, said the editors, Paul is "a genuine American folk character, created by the people themselves, in the bunkhouses and ordinary logging camp." Such a claim has been made many times, before and since. Paul is said to have grown out of the American environment by some folksy process, a pure product of mass creativity. In his preface to a Bunyan collection by Harold Felton, Carl Sandburg asks who made Paul Bunyan and answers: "The people, the bookless people, they made Paul and had him alive long before he got into the books for those who read. Paul is as old as the hills, young as the alphabet."

The giant lumberjack described by *Fortune* and Felton was not born in a bunkhouse, but in an advertising office. He was considerably younger than the hills; no older than a promotion scheme of the Red River Lumber Company; not as old as Sandburg's *Chicago Poems* (1916).

This is not to say Paul never was a lumber hero. Certain old loggers talked about him before the office crews wrote about him. Their oral yarns didn't concern the universal roustabout but a vocational in-group hero whose talk and feats could appeal only to those on the inside of a highly specialized business. These jargon-filled tales, mainly concerned with exaggerated adversities and tender tyros, were funny only to be initiated. When times and methods changed, and the need for brute strength and endurance passed, the oral Bunyan tradition began to die out too.

The few authentic Bunyan camp stories recorded before World War I were supplied by men who were old even then.

The Bunyan known to most Americans—he might be called Paul Bunyan II, to distinguish him from Paul Bunyan I, the lumberjack's bona fide hero—is a giant of a considerably different order. He has a literary, not a folk, origin. He speaks "printed page" English and not vernacular English, and never gets technical or profane. He utilizes parachutes, steam-driven concrete mixers, pipe lines, and other devices unknown before the turn of the century; and he ranges far afield to work in mines, airfields, and ranches, but never as one group's hero. His exploits are not passed on in ballad or song. Nor do workmen improvise on the old themes at their leisure, or invent new situations. Most lumberjacks do not read the literary accounts where the real improvisation takes place. Today Paul Bunyan serves a multitude of functions and ends that the creators of Paul I never dreamed of. For Paul II is no mere comic hero. On the contrary, he is an exponent of American know-how, industrialism, and the "American way of life."

The person responsible for Paul II was a young freelance advertising man named William Laughead. He had worked for eight years in Minnesota logging camps, moving up from choreboy to assistant cook and on to timber cruiser, surveyor, and construction engineer. When he went into the advertising business he hoped to put his logging lore to good use; this assignment was almost too good to be true. Right away he remembered some funny stories about an imaginary lumberman which might make good copy if they were put into a palatable form. The name usually given this character, Laughead recalled, was Paul—Paul Bunyan.

Laughead says he first heard of his hero-to-be from some old loggers in Minnesota. He sensed that Paul would make a good trademark for a lumber company. His antics were distinctly colorful; except for Malloch's poem he had not appeared in print; and, for all practical purposes, he was totally unknown to his fellow Americans. Of course, the jargon-filled and frequently profane Bunyan stories the old-timers had told were unsuitable for a general reading in public, but Laughead saw distinct possibilities.

After a few months he produced a booklet of thirty-two pages called "Introducing Mr. Paul Bunyan of Westwood, California." It was divided equally between Bunyan material and straight advertising. A "warning" on page one indicated that Laughead intended to pass Paul off as a widely known folk hero and a paragon worthy of taking care of Red River Lumber Company problems:

> Everything we tell you about the Red River lumber and its manufacture is the Gospel Truth. Everything we tell you about Paul Bunyan is lumberjack mythology. Paul is the legendary hero whose exploits have been related to generations of tenderfeet

from time immemorial. We have gathered some of them up, and
pass them on to you with a few trimmings of our own.

It would be very interesting to know just what those "few trimmings" were; Laughead is not sure where recollection ended and con-
trivance began. For inventing such Bunyanesque characters as Babe
the Blue Ox, Brimstone Bill, and Big Joe the Cook, he takes full
credit. No one disputes his claim that he first vested Bunyan with a
commercial mission. He was personally responsible for devising the
first pictorial representation of Paul Bunyan (his Paul has a round
face, black hair, and a cat's-whisker moustache), which was copy-
righted as the Red River Lumber Company trademark. Accompany-
ing the original portrayal is a reminder that "he stands for the quality
and service you have the right to expect from Paul Bunyan." Under
the picture is the warning that the picture is "registered." It is as clear
a case of *ars gratia pecuniae* (art for money) as one will ever find in
the annals of American popular heroes.

After a slow start the Bunyan booklets picked up speed and gained
wide circulation in a few years. In 1922, when Laughead was made a
regular employee of the Red River Company, he prepared a large
pamphlet which was mailed not only to potential customers but also
to newspaper editors, historians, and the general public. All found the
Bunyan stories to their liking. For the editors they made good copy;
for the historians, grounds for endless speculation and quibbling;
and, for the general public, items that inflated national and regional
egos.

The lumber states knew a good thing when they saw it. They decided that Bunyan was a latter-day epic hero who had been roaming
around in their forests since they were planted. Minnesota became
belligerent over Bunyan, and possessive about all stories or legends
that surrounded him. Neighboring states Michigan and Wisconsin
fought to center Bunyan lore within their boundaries. Paul and his
blue ox bedecked the walls of the new student union building at the
University of Wisconsin, while Michigan supporters proclaimed that
the "Saginaw Paul Bunyan" was the real thing. Far Western states entered the competition with enthusiasm; in recent decades Oregon has
become one of the leading promoters of Bunyania.

Soon scholars joined copywriters employed by tourist centers,
chambers of commerce, state agencies, and booster clubs in the campaign to win national recognition for Paul. Many devised methods to

prove that he was indeed a folk hero, a kind of Beowulf west of Boston; others explained why the lumberjacks had decided that Babe the Blue Ox was precisely forty-two axe handles long, and a plug of Star tobacco between the eyes. A somewhat hardbitten scholarly segment, finding more of the hoax than the historical in Paul, squared off for wordy battles with the faithful.

America was ready for Paul Bunyan. High-powered advertising and journalism needed copy. Historical and literary scholars and, more particularly, the rising school of folklorists needed subjects to investigate. Painters and poets were searching for new native themes. Chauvinism gave a nudge and Paul Bunyan filled a cultural vacuum.

Bunyan promulgators did a good job. Their candidate is now generally regarded as the leading American folk hero. So far as organizational backing is concerned, the Red River Lumber Company remains Paul's best friend; it has distributed 100,000 Bunyan pamphlets and, incidentally, a good deal of wood. Finally, the firm sold all its lumber, and the same Archie D. Walker who put Laughead to work creating Bunyan in 1914 helped to liquidate the company's holdings in 1945. There is something characteristically American about this. In the lifetime of a single man much of the timber of two heavily wooded areas had been cut with little thought of the morrow. The stories fabricated to sell this lumber were reaching their greatest popularity just as all the Red River Lumber had been marketed. Whether that gentler and in many ways more appealing folk hero, Johnny Appleseed, will be able to keep up with Bunyan is a question folklorists have not been able to answer. For some reason the hero who planted trees has not fared so well as the one who cut them down.

Meanwhile other writers had not waited for the liquidation of the Red River Lumber Company to make use of its symbol, "registered" or not. In the 1920s, newspapers in both Seattle and Portland featured tall tales about Paul Bunyan supplied by their readers. Another lumber company publicist, James Stevens, has become nearly as important as Laughead in popularizing the legend. In 1925 he produced the first group of Bunyan stories to be published by a New York press. This volume would be later supplemented by several articles in popular magazines, a second edition of the book, and a willingness to do battle with anyone who dared to challenge the folk authenticity of Paul Bunyan II.

Whatever Stevens' merits as a historian or folklorist, he is a gifted writer and devout worshiper at the axe-hewn Bunyan altar he has so diligently helped to build. Originally his interest in Paul was purely professional. He was director of public relations for the West Coast Lumberman's Association when he first learned of Laughead's success.

Bunyan booster Stevens gave the giant a new elegance. His Paul was "powerful as Hercules, indomitable as Spartacus, bellowing like a furious Titan, raging among the Queen's troops like Samson among the Philistines." The lumbermen who raided the forests were "toiling demigods and sweating heroes" of the dark woods, workaday poets who created "the marvelous mythical logging camp, with its cookhouse of mountainous size and history of Olympian feats." Though Stevens demonstrates to his own satisfaction that the historical Bunyan was a Frenchman and though he constantly utilizes the framework of the classical hero in his own stories, he is sure Paul Bunyan, as he stands today, is absolutely American from head to foot; that thousands of lumberjacks contributed to the classical picture of Bunyan, who will live as long as there is a forest for his refuge, as long as there are shadows and whispers of trees.

Though Stevens' book was a popular success, it had its detractors. J. Frank Dobie, Texas scholar and folklorist, noted in a review that Stevens assumes the liberties of a judicial novelist, frequently putting his own dull ideas into the mouths of his characters, even of the great Paul himself. Dobie found some sections utterly foreign to the bunkhouse yarn.

A more thorough attack on Paul II was launched by Carleton C. Ames in his article "Paul Bunyan—Myth or Hoax?" in *Minnesota History,* March 1940. While others merely speculate about whether the old lumberman really knew about Paul, Ames took to the field to find out. Touring northern Minnesota and the drainage basins of the Chippewa and St. Croix rivers, he interviewed hundreds of lumbermen, with startling results. Not one of them had ever heard of Paul Bunyan. It seems incredible, concluded Ames, "that if Paul Bunyan yarns had any kind of circulation during the heyday of the loggers, they should not have come to the attention of at least some of these individuals."

Cries of defamer and debunker were showered on the young investigator from all sides. Newspaper editors peppered their editorial pages with anti-Ames tirades; there was even a suggestion that the state legislature should investigate the University of Minnesota. But other scholars came to Ames's defense. Mrs. Grace S. McClure, State Librarian of Michigan and a native of the Saginaw region which was supposed to be a Bunyan stronghold, stated that in hundreds of talks with lumbermen she had never found a single one who knew about Paul Bunyan, and that she didn't even know of the literary accounts in Michigan until around 1930.

Stevens had a chance to reply when he wrote an introduction for a second edition of his Bunyan collection in 1948. In it he castigated the professors who had "sweated their pale blood to prove the legend a hoax, only to expose themselves as futile dabblers in nonsense." He also revealed that for three years he and his wife gave time and income to the pursuit of clues and the study of wearisome old lumber trade publications in the hope of finding some bit of documentary evidence to prove that Bunyan was older than the pure-blooded scholars intimated. Nothing was found. The Paul Bunyan that Mr. and Mrs. Stevens were defending and the disappearing oral hero of the lumberjacks were different creatures, with different origins and histories, brought into being for different purposes.

Yet the "genuine folk hero" had caught on. He was prominent at both the New York and California World Fairs in 1939. He has made his appearance in ballet, opera, painting, and poetry; artists as different as Robert Frost, Carl Sandburg, W. H. Auden, and Benjamin Britten have celebrated his exploits in various ways. Books about him continued to appear.

Bunyan benefits from the adaptable American hero pattern from which Mike Fink, Kit Carson, Davy Crockett, and Daniel Boone are cut. Paul is an oversized Boone, sharing Carson's mobility and Fink's colorful vocabulary, and improving on Crockett's extraordinary feats. Of course, Paul has more "know-how" than any of his nineteenth-century predecessors. With the shift from an agrarian to an industrial society, one might predict that there would be a shift from heroes who use their hands to heroes who use machines. So it is with Paul II. Robust, uncouth, destructive, resourceful, friendly, when he moves into the forest he gets things organized and he gets results; he can't be very much concerned with who or what suffers in the process. Where others think in terms of tens or hundreds, he thinks in terms of thousands. He believes in the survival of the fittest.

W. H. Auden suggests that in essence Bunyan is a bumptious, swaggering, optimistic nineteenth-century Victorian, a projection of the collective state of mind of a people whose tasks were primarily the physical mastery of nature. Paul II of the advertisements does not subdue nature (as Boone and Crockett did) with his physical strength alone. He depends on the latest technology and machines, and enjoys nothing more than a flowing pipeline. He is obsessed with the problem of saving time. He devises methods of doing things faster and with less effort. In short, Paul Bunyan has brought the assembly line into American folklore. He stands for efficiency plus.

In an age that enjoyed unprecedented business expansion and prosperity, Paul Bunyan became ever more prominent. As new industries and factories are opened, advertising agencies have put him to work and sent him to fight. Back in 1941 Bunyan went to war with the greatest of ease. It was a war of production and procurement, and in these matters Paul was a past master. Hadn't he had boys skate around his skillet with bacon on their feet to grease it for his pancakes? Hadn't he installed elevators to carry pancakes to the ends of the table, where boys on bicycles rode back and forth on a path down the center of the table dropping cakes when called for? Hadn't he fed his men pea soup through a pipeline, and perfected a method for pulling curves out of rivers?

In a century in which virtual reality is driving reality off the scene, Paul Bunyan seems to fit in very well.

Ever since Henry Ford put America on wheels in 1908 with the Model T, Americans have craved and demanded new automobiles. What to do when new models come out? Junk the old ones. Thus does nature become technology's junkyard.

The Sign of the T: Henry Ford

"The Case of Little Reuben occurred only twenty-three years after Our Ford's first T-Model was put on the market." (Here the Director made a sign of the T on his stomach and all the students reverently followed suit.)

Aldous Huxley
Brave New World

History is bunk. All the art in the world isn't worth five cents. Reading books musses up your mind. Literature is all right, but it doesn't mean much. The jagged crystals of sugar damage your stomach. Cows, horses, and creeds will disappear from the earth. Don't ruin a son by sending him to college. Anyone who smokes or drinks shouldn't be employed. The theory of evolution is part of the Jewish plot against the Gentile world. So was the assassination of Abraham Lincoln.

These statements and dozens like them were all made publicly by an American hero. He was no backwoods bigot or mere theorizer, but one of the most respected and practical men of our time, the admired billionaire. People who never heard of Washington, Lincoln, or the Declaration of Independence knew his name as well. It bounced over the world's highways daily. The name was Henry Ford[1] (1863-1947). He put us on wheels. Where are they taking us in the twenty-first century?

Ford had, as much as anyone of his times, a sense of mission. "Machinery is the new Messiah," he claimed. If this be true Henry Ford is a saint and his blueprints are Holy Writ. That our offspring will make the sign of the T and measure T A.F. (After Ford), as Aldous Huxley mischievously suggests, is unlikely, but they may well revere him as a miracle worker who assembled a brave new mechanical world. Outside his factory, he lacked heroic dimensions, but he achieved immortality as the Great Mechanic, the Know-How King.

Ford had no sympathy for intellectuals and dreamers. Nor had he any reason to think that as a diplomat, historian, or sociologist he had succeeded, despite his strenuous efforts in these callings. He loved the factory, gloried in its productivity, and exulted in being a mechanic. For forty years he depended on his own tools for a living. After that he employed a mechanical army without peer or precedent. His life was a progression from watches to horseless carriages to Model T to Model A to V-8 to B-24.[2]

At Ford's the mechanic was Superman. Because "Cast Iron Charlie" Sorenson made machines and assembly lines fill incredible quotas, he was the most admired employee Ford ever had. After awarding him a six-figure salary he reportedly added, "If that isn't enough, Charlie, take some more." Even world wars raised problems that for Ford called for mechanical solutions. "It is my earnest hope to create machinery to which those who so desire can turn to inquire what can be done to establish peace," he told the press when his peace ship sailed.

My Philosophy of Industry (1929) showed Henry Ford to be still a mechanic dabbling in sociology. Just as accurate gauges and methods produce a smooth-working, efficient machine, so clear thinking, clean living, square dealing made human rehabilitation sound simple. We should be able to renew our human bodies as we mend a defect in a boiler. "Machinery," he wrote, "is accomplishing in the world what man has failed to do by preaching propaganda, or the written word. Political opinions don't really make much difference." To him morality was "merely doing the sound thing in the best way." Ford's followers in Aldous Huxley's *Brave New World* quite appropriately made the sign of the T on their stomachs, not on their heads or hearts.

Of Scotch-Irish lineage, Henry Ford was born in rural Michigan in 1863. Hating farm chores, he spent his spare hours fixing farm machinery and watches. At sixteen he overrode his family's wishes and went to Detroit. To appease his parents he made a second short try at farming, after which he married and returned to Detroit. As an Edison Company employee he rose from night fireman to chief engineer. Then he quit, to spend all his time marketing the gasoline quadricycle he perfected in 1896. He moved effortlessly into the infant automobile industry. In 1903 he set up the Ford Motor Company with a capital investment of $28,000. Five years later the first Model T came from his factory; five years after that his assembly line was a reality.

In 1914 he established a minimum wage of $5 a day in his plant and was skyrocketed out of obscurity. His rugged Model T became America's leading car and the subject of a unique folklore. Before the Great Depression he had sold 15,000,000 tin Lizzies and became a billionaire.

Ford's later career was anticlimactical. The Model A and V-8 cars did not maintain the Model T's supremacy. Labor and plant difficulties multiplied; the son who had held the title but not the powers of president of the company died in 1943. Henry remained in control of the empire he had created until senility and death overtook him in 1947. Over 100,000 Americans from all walks of life filed by his casket for a last look at the mechanic almost no one called Henry to his face. The man was gone, but his name and his cars rolled into history.

Ford's workaday and respectable rise lacks the color and drama that went with Phineas Barnum, Diamond Jim Brady, and Jubilee Jim Fiske. He reflected the strength of the rural America which sustained his legend, but this was insufficient grounds for his status. We must go deeper to find his unique strength. Whimsical and retiring, he was at his worst in public appearances. Few admired him, and even fewer could claim to have been his friend. Most workmen thought of him as a spy and slavedriver; his legend did not spring, like the Lizzies, from the assembly line. The Mechanical Wizard was more esteemed by those who drove Fords than by those who made them. "To get along with him," an employee reminisced, "you had to have a little mean streak in your system."

During New Deal days Ford's reputation went down; when his pro-Nazi tendencies became known, it hit rock bottom. After his death Ford's fame underwent the hysteria of depreciation that frequently follows a hero's death. In 1953, during a highly publicized fiftieth anniversary of the Ford Motor firm, this trend was reversed. His heirs dramatically opened Ford's private papers to the public; although he had said history was bunk, he had collected enough personalia to cover 5,000 feet of shelves in the new Ford library.

Life magazine suggested that we take "A New Look at Ford" and devoted fourteen pages to improving his national standing. Other magazines followed suit. Company advertisements took advantage of the technological growth that followed World War II; Henry Ford II worked hard to improve public relations.

Perhaps the time is ripe to note objectively just what the living Henry Ford's attitudes were, insofar as we can piece them together from existing documents and memories.

Throughout his long life Ford suspected social innovation, criticism, and expert opinion. A dissenting voice infuriated him. He fired so many people that a special name was finally coined for the unfortunates—the Ford Alumni Association. Once he decided to replace all physicians at the Henry Ford Hospital with chiropractors; another time he cut off milk deliveries at Greenfield because medical men would not say that his son Edsel had contracted undulant fever. "Never trust ear-piddlers or lawyers," he said. His personal world was as superstitious as his factory world was scientific. Fear of Jews, John Barleycorn, the du Ponts, and international bankers would not let him rest. Black cats, broken mirrors, omens, and coincidences terrified him. Never would he change a sock that had been put on inside out. Not impressed with church theology, he nonetheless believed in reincarnation. This is how he reasoned it: "When the automobile was new and one of them came down the road, a chicken would run straight for home, and usually get killed. But today when a car comes along, a chicken will run for the nearest side of the road. That chicken has been hit in the ass in a previous life."

The mind of the Automobile King roared down the highway of night without headlights. So affluent that when he crashed through the barricades no one chastised him, he listened so seldom that he learned little from other drivers. On he rolled, a hit-and-run driver whose victim might be the Jewish race, the Catholic religion, labor unions, leisure, the New Deal, or Wall Street. Ford's prejudices were those of rural midwestern America. William Jennings Bryan, that crusading enemy of city slickers and religious modernists, supplied him with his political credo. Like Bryan, whom he befriended, Ford was a pacifist and isolationist. When newspapermen came to the docks to cover the departure of the Ford peace ship for Europe, it was Bryan who posed with Ford on deck.

"Any customer can have a car of any color he wants, provided it's black," he told his advertising staff. Only after a public rebellion reduced sales would he admit that he was not the final arbiter on how an automobile should look. As his dogmatism increased with his assets, so did his bitter opposition to unions. The Ford Motor Company was run like a feudal dukedom, and governed, said *Fortune* magazine in

1941, by a mutually distrustful group of executives, "most of them without titles, with no clear lines of authority or responsibility anywhere delineated." If he trusted few men, he idolized the one who encouraged him to build automobiles. This was the tinkering Thomas A. Edison, whose relics Ford collected in a special museum, and whose front yard Ford caused to be dug up and carried reverently to Dearborn. He even kept Edison's dying breath in a bottle.

Both Ford and Edison were cut from the same extraordinary bolt of cloth. Considering how highly we prize the gadgets and artifacts with which their two lives were obsessed, it is small wonder indeed that America has lionized them. They were the Rover boys on the Trail of New Trinkets.

An episode from Ford's early career accents his Edison idolatry. In 1914 he hired a special train to make the Detroit-Port Huron run on which Edison had worked as a boy. He purchased a vendor's box, and had Edison sell once more his nickel treats. Then, at the end of the run, Ford arranged to have his guest send a telegram from the station where he had served his apprenticeship. The aging Edison ticked out a halting message to his son in New Jersey. Delayed by Ford's historical pageantry, a distant operator broke with, "Tell that kid to get off the line." He was right. There was a kid on the line, and another kid had arranged his being there. Ford remained, in all but mechanical matters, a child. Like his own Model Ts he was a triumph of functionalism, made to go but not to ponder.

Once he got going, Ford was hard to stop. Under him burned the fires of John Calvin's hell. His strict Scotch-Irish rearing left a deep mark on him, which wealth and leisure could never conceal. Engraved over his fireplace was Franklin's line, "Chop your own wood and it will warm you twice." Laziness was at the bottom of America's troubles. The Bible verses he memorized dealt with hard work, frugality, and justification by works. Rich Henry is, among American heroes, merely a mechanized version of Poor Richard.

If God gave John D. Rockefeller his dimes, he must also have given Ford his dies. If men would only listen to him and do his bidding, thought the Motorman, he would run them as effectively as he did his factories. "When I think about the thousands of families dependent upon my enterprise," he said, "there seems to me to be something sacred about the Ford Motor Company." Recent critics have found so much wrong with Ford's methods and policies that they have overlooked the lofty ends. The money he made neither spoiled nor satisfied him. In the best Calvinistic tradition, he considered himself his brother's keeper whether his brother liked it or not. Not to remove the stain from his workmen's fingers, but from their souls, did he outlaw smoking in his plants. When he announced he would pay a minimum wage of $5 a day, he took it for granted that only moral people deserved such a rate. He assumed that he could manage men as he could manipulate machines. That he never fully understood the social effects of his actions should not make us forget that he never lost his concern about them. Aware that his innovations changed millions of lives, he was determined to deal with the human consequences. Even at his worst he was only trying to graft a new branch onto the tree to rural American life. No other industrialist in his era made so extensive, or so unsuccessful, an effort.

Though America's leading industrial figure, Ford was unlike any of the other heroes we have examined. His personality was not pleasing, and his utopian dream was unsuccessful. Yet he became, while still alive, a popular folk hero. This cannot be explained by saying the people were wrong about the real Henry; instead, they invented a legendary Henry who had a Model T brain. As with that contrived folk hero, Paul Bunyan, there is really a Henry A and Henry B, quite different in history and character. Such a division has been the people's prerogative for centuries, and personalities differing as widely as Ulysses, Roland, Robin Hood, Billy the Kid, and Eva Peron demonstrate it. Both Henrys are oversimplifications of capitalism. In fact and fancy Henry was a farm boy, of humble parentage: a tinkering dreamer who didn't mind putting his visions to the test; a person whose natural ability carried him forward in a highly competitive field; a mechanical genius in a mechanized age. A major need of his day was cheap transportation. He provided the answer, on a scale no man before him had even visualized. Here the two Henrys split, Henry A becoming a tyrannical motorman, and Henry B the guiding spirit in a culture which has used its productivity as a chief means of survival. Henry B is as fortunate as Henry A is pathetic. Legend makers have succeeded in dramatizing early Ford stories, such as his first successful run in a gasoline buggy during his early manhood. They have made as much of the last hours of preparation as of the hours when Lindbergh was defying the lonely Atlantic, or Lincoln was awaiting word of the chance meeting of the Northern and Southern armies at Gettysburg. Ford worked around the clock until, at two in the morning, he was ready for the final test. Rain drenched the muddy Detroit streets, but he didn't even notice it. Would the contraption run? Mrs. Ford put a cloak over her shoulders and went ahead of him with a lantern. For a terrible moment nothing happened. Then Ford found that a screw had come off; he replaced it. The thing ran. The automobile age had begun. Here was drama that the man on the street could understand.

Even when Henry Ford became a billionaire his hands stayed clean, for no man had been robbed. Where no business had existed he had created one. By 1940 one out of every seven Americans was employed in the motor car or allied business. Ford remade transport and put the Horatio Alger myth on wheels. Clerks and farmers entrusted him with the $100 tucked away in the mattress; Henry was on their side. Hadn't all the giant corporations called him crazy, and hadn't *The Wall Street Journal* attacked him? He'd made Detroit in 1914 what California had been in 1849: the end of the rainbow.

The Ford five-dollar minimum day came with a number of strings attached. Only workmen who were strong qualified. Married men had to be living with and supporting their family; single men over twenty-two had to be living "wholesomely" and display "proud thrifty habits"; men under twenty-two and all women had to be the sole support of their next of kin. Henry was as anxious to keep vice down as he was to keep production up. Rural America, fearing the sins of the city, thought it a fine idea. The just would be rewarded, and the unjust cast out. That the reward, at least monetarily, averaged less than $1,000 a year per worker in 1930 didn't destroy the illusion.

Eventually the Motor King became a touchstone for testing proposed changes and laws. Ford's was the American system, and anything that hurt him hurt America. In a culture that had made the dollar an almighty thing, Ford collected a billion. Yet, said the legend, he was as simple and sturdy as when he plowed his cornfields or tinkered in his tiny garage. This popular conception of Ford collided with certain historical facts, but overran them and rolled on. It was hard to explain how a humanitarian could allow cold water to be turned on shivering unemployed workmen, or have police spies to watch all employees. People outside Detroit knew little of these occurrences. It was hard to reconcile Ford's alleged simplicity with his 100,000 acre Georgia plantation, fashionable London home, private railroad car, and million-dollar Dearborn estate complete with a $30,000 pipe organ, $37,000 boathouse, and $69,000 garage. Ford's legend makers answered with a counteroffensive. Didn't Henry love the old-time square dances and country hikes? Wasn't he still a dirt farmer at heart? Hadn't he borrowed two cents to buy the first Edison commemorative stamp, being out of change at the time? Hadn't he foiled the Wall Street bankers who tried to destroy him? These touches reassured the public. The tin Lizzies served not only to transport Americans, but to foster tales, slogans, and jokes about the man who was responsible for their existence.

Lizzie-lore is a key to understanding Ford's reputation. Early Model Ts, objects of awe and delight, symbolized what Ford could do and what America would become. Ugly but useful, Lizzie came just in time to create a suitable basis for a new machine age folklore. No mechanical object has ever been the butt of so many jokes, or source of so many wisecracks as the Model T, alias Leaping Lizzie, Little Bo-Creep, Rolls Rough, Wanderer of the Waistland, the Answer to a Walking Maiden's Prayer, Bouncing Betty, Graf Zep's Uncle, Lizzie of the Valley, Passion Pot, and variations beyond number. Folklorist B. A. Botkin collected over a thousand Model T nicknames for a piece in *American Speech.* Some, such as Flapper, No Charleston, and Wanted—a Bootlegger to Share Our Home, were dated by the 1920s. Some, like Henrietta Elizabeth Van Flivver or September Morn, affected the grand style; others moved over into Freudian realms. Consider, for example, the label, "I'm a second son of a last year's Lizzie," accompanied by a pair of coconuts hung from the differential.

The labels stressed Lizzie's all-around cussedness, and capitalized on America's fondness for slogans and wisecracks. They were mechanical adaptations of the mule and burro stories of European folklore, redone in the American idiom. When you saw a car bumping along labeled, I'm from Texas, You Can't Steer Me; Columbus Took a Chance—Why Can't You?; or Sugar, Here's Your Daddy, you knew what country you were in. And you knew frontier optimism wasn't really dead when others read, We Ain't Climbed Pike's Peak Yet, or Lazy Loping Lizzie on her Last Lengthy Leap. Lizzie's maker also figured in all the buffoonery. His staff called him Mr. Ford, but his customers preferred Henry. Flivvers were inscribed with such lines as, Turn over, Sis, Henry's Here, or Don't Worry, Lizzie, Henry Will

Fix You Up. Behind all the razzing was pride in Ford's achievements. "When better Buicks are built, Fords will pass them," proclaimed a popular slogan. Even Lizzie's obvious failings, such as a tendency to drop vitals en route, came in handy. Follow Me for Ford Parts was a favorite.

Ford jokes flourished like Ford labels, both orally and in printed joke books. The usual theme was confidence in the man and in the cars that were transforming America. They stressed inventiveness, productivity, or some ingenious triumph at the Ford plant. After-dinner speakers told about the old lady who sent her tomato cans, or the farmer on his tin roof, to Detroit, and got a Ford car back by return mail. Rural humorists poked fun at the Eastern banker who paid $15,000 for a fancy French limousine but kept a Model T in the trunk to pull it out of mud holes. They told the sad tale of the Ford worker who dropped his wrench on the assembly line and was twenty cars behind after he bent over and picked it up. Few of the jokesters knew it, but their tales were cast in the mold that had served for centuries. The Ford joke was the drugstore chronicle of the new gasoline age. Neither Jove nor Charlemagne nor Beowulf had done anything more remarkable than Henry Ford, who turned bits of tin into automobiles.

Had you heard that next year's Lizzies were going to be yellow so they could be sold in bunches like bananas? That Ford planned to paper the sky with flivver planes? That Model T's were being shipped in asbestos crates, since they came off the assembly line so fast they were still hot and smoking? That a man got rich following Ford's cars and picking up the parts that fell on the road? That Uncle Jeb got shaken up in Lizzie until the fillings came out of his teeth? That the next Ford would come with a can opener, so you could put doors anywhere you wanted them? That Henry would pay $100 for any flivver joke which made him laugh?

Though not as frequent as they once were, Ford stores still thrive on the American scene. In May 1953, Art Thomas, owner of a 1908 Model T, was fined for speeding. His courtroom comment was in keeping with the Lizzie tradition; "It was only hitting on three. If it had been hitting on all four I doubt if they would have caught me." On and on the jokes go, enhancing the reputation of a man few Americans saw or heard. They keep Ford in the realm of heroic immortality.

Lizzie brought golden days to Ford. Introduced at the 1907 Chicago Auto Show, the Model T went immediately into production.

Henry offered a stripped car. Everything was standard and inter-changeable, which made for simplicity of operation and repair. The end product was the first workable solution to the problem of putting a democracy on wheels. As the 1920s rolled along on sturdy Ford axles, politicians even began to talk of two cars in every garage. Henry never endorsed such extravagant notions; one would do, if it were a Ford. In two decades he produced 15,456,868 Model T cars in thirty assembly plants, selling them for as little as $265 apiece. When he started making Model Ts his surplus balance was $2,000,000. Before he changed to a new engine it was $673,000,000. Although by 1923 he had not spent a penny on advertising for five years, he was making 7,000 cars a day and still not filling all orders. Ford seemed to have answered the criticism about capitalistic production methods, and the vitality of laissez faire. Instead of condemning what Ford did, the Russians tried desperately to imitate it. Europeans poked fun at Ford's methods and products, but envied Americans their Ford cars. A British writer predicted that Americans would eventually give up their homes to live, wed, and die in their Lizzies.

By 1925, however, Ford sales were declining. The public rebelled against Lizzie's absence of comfort, beauty, and style. Chevrolets, Reos, Overlands, Dodges, and Buicks improved until they matched Lizzie in performance, and surpassed her in style and comfort. Ford refused to change. Reluctantly he agreed to an elaborate advertising campaign to win back Lizzie's buyers. Falling sales continued, show-ing that even in Lizzies there was a saturation point. Finally the Ford plant was converted to another standardized engine, the Model A, which lasted from 1927 to 1931. Again the public balked. A disillu-sioned Ford finally acquiesced to a yearly model and to the trimmings he so detested. Stories continued about Henry. Many of them can be attributed to the clever and able men whom Ford hired to promulgate the Ford image. His assembly-line chief, Charles Sorenson; his pub-lic spokesman, William Cameron; and his personal factotum, Henry Bennett were instrumental in making him a major culture hero. The postwar success of his grandson, Henry Ford II, rejuvenated that reputation in the 1950s.

In his later years Ford became an antiquarian, spending thousands of dollars to preserve artifacts and documents of the way of life his mechanical improvements had helped destroy. This pious turning to the past fascinated Americans. Rumors about his projects swelled the

Ford folklore. Had you heard that Ford wanted to ship the Florida Everglades to his museum? That he was going to buy Plymouth Rock, the French Navy, the Russian Crown jewels? Did you know he wanted to move the Great Wall of China? Would Henry give a Ford car to any girl who didn't bob her hair, or to any woman who never wore knickers? Such questions pointed up a Ford trait that endeared him to America. When Henry decided to make a public gesture or statement, he generally ended by falling flat on his face. His social theories wouldn't work, his prejudices were untenable, his political hunches were unfounded. The Ford blunders, many of which made the headlines, showed just how fallible this disciple of industrial infallibility really was. When the public decides a man is a snob (as it did about Hamilton, Van Buren, and Dewey) it will never give him its heart. Whatever else Henry was, he was not a snob.

That Henry Ford was quite apt to be as wrong as the next fellow was best illustrated by the chartering of the ill-fated 1917 peace ship which was to "get boys out of the trenches by Christmas." Persuaded by a Hungarian named Rosika Schwimmer that a Ford ship would bring Europe to its senses, Ford chartered the "Oscar II" for the job. Commenting on Ford's decision, John Dos Passos says (in *The Big Money*): "Any intelligent American mechanic could see that if the Europeans hadn't been a lot of ignorant underpaid foreigners who drank, smoked, were loose about women and wasteful in the methods of production, the war could never have happened."[3] So the Great Mechanic embarked, having authorized the sending of a cablegram to Pope Benedict VII who, alas, had died in 983 A.D. The American press gave the event as much space as to President Woodrow Wilson. Reporters had a field day, describing at length the reformers, utopians, and crackpots who went over for the trip. All state governors were invited, but only the governor of North Dakota sailed. Ford's old friend William Jennings Bryan went. A picture of him holding a squirrel cage on deck, arranged by a clever photographer, made the rounds of American newspapers. The climax came when the bizarre Mr. Zero, flophouse reformer, jumped into the North River and swam after the peace ship, which had sailed off before he arrived.

Henry Ford was not long in perceiving that his efforts weren't going to empty many trenches by Christmas. Claiming to be sick, he returned as quickly as possible. Reporters met to ask gleefully whether he had made any important discoveries about Europe. Henry, who never let a social blunder get in the way of his business eye, said yes. While in Europe he'd discovered there was a seller's market for tractors in Russia. Later on, Soviet orders helped sustain Ford during the Depression.

That Ford could claim history was bunk and still set up historical museums created a newsworthy paradox. When in 1919 he sued the *Chicago Tribune* for a million dollars on a libel charge, the defense used the occasion to expose Ford's ignorance of history before the world. Ford was made to admit he didn't know there had been an American revolution in 1776, and that he thought Benedict Arnold was a writer. Rather than read aloud in court, he admitted he preferred to leave the impression that he could not read at all. Ford was awarded six cents in damages. The public got a million dollars worth of chuckles. In 1934 Ford announced that he had purchased the authentic Pittsburgh birthplace of Stephen Foster. When competent historians showed he had been the victim of a hoax, he merely bought "sworn statements" to support his claim. He thought the plans to reconstruct Williamsburg, Virginia, were silly and stated this publicly. Later the Rockefeller family turned the early capital into an important tourist center. Ancient history was Ford's weakest suit. He maintained that cars and airplanes had existed before written records were kept. All in all, he was as out of place in a history discussion as a prairie schooner would have been on the Ford assembly line.

Some Ford blunders were quite harmless; others were not. At various times he launched crusades against tobacco, liquor, unions, Jews, Catholics, international bankers, and anti-Nazi statesmen. Articles from *The International Jew* ran in the *Dearborn Independent* for twenty weeks. In each issue was printed the claim that, "The statements offered in this series are never made without the strictest and fullest proof." Four volumes of anti-Jewish writings were issued by his Dearborn Publishing Company in lots of 200,000. Copies were particularly prized in Nazi Germany. When sales sagged, Ford recanted his anti-Jewish stand; his claim that he never knew the articles were being published was nonsense.

Deeply isolationist and anti-British, Ford flaunted his pro-Hitler sentiments for years. In 1938 he accepted the Award of the Grand Cross of the German Eagle from the Third Reich, and was photographed with the cross pinned on a specially made tuxedo. "I don't know Hitler personally," he told the *Detroit Press* in 1939, "but at least Germany keeps its people at work." Fritz Kuhn, leader of the German-American bund, was on Ford's payroll; William Pelley, Silver Shirt leader, and Gerald K. Smith, professional bigot, also received Ford aid.

But the people forgot how confused Henry had been in politics when he promised to turn out a thousand planes a day to lick the Axis. He never did so, although Charles Sorenson eventually produced many B-24 bombers in the Willow Run factory. What Henry the mechanic did always offset what Henry the blunderer tried and couldn't do. His crotchety integrity and wry petulance delighted rural America. His short, crabby answers matched those of Calvin Coolidge. Indeed, the lank figure of Henry Ford could easily pass for Uncle Sam. There was no frustration or regret mirrored on his lean face. Public figures came and went, but Ford kept producing cheap cars.

Although it had its comic moments, Ford's career when viewed as a whole was a tragic one. Endings of heroes' stories are almost always tragic. So much sorrow abounds in common life that we look for and expect it in the lives of the great; from this we obtain major comfort. In what sense was billionaire Ford tragic? He put cars and filling stations on every road, but kept thinking in horse-and-buggy terms. He devoted his life to perfecting a technology for the future, but spent his last years among exploded delusions of the past. His life was associated with machines, but he longed to dance around the village maypole or sun himself on the porch of the Wayside Inn.

Realizing that he had created problems for the twentieth century, Ford tried desperately to solve them—always in terms of the eighteenth or nineteenth century. He never perceived what he had actually created. Charles Chaplin's masterful burlesque of the assembly line in the movie *Modern Times* could not have amused Ford, who intended to free men's minds by the very devices Chaplin ridiculed. At River Rouge, just as on Brook Farm, Ford expected the workman to begin his real day after his routine work was over. His error came in confusing means with ends. The River Rouge plant proved, as Chaplin's movie implied, that a world in which production becomes an end in itself is as inhuman as the machines.

The men on the assembly line could have told the mechanical wizard that he had perfected the cycle of production but not the lives of those who produced. Sameness and tedium ate into their souls. Night foremen would tell of men who worked efficiently though sound asleep, muscles moving but brains dormant. A Ford worker was not allowed to speak to a pal, or break the monotonous rhythm of the line. If a machine stopped, a bell rang automatically and a foreman came running. The joke about the workman who dropped his wrench and got twenty cars behind wasn't so funny when you knew that as a result he was fired. An employee couldn't even go to the toilet unless a substitute was available to fill in. No wonder men cursed Henry Ford in sixteen different languages. If all the pent-up hatred had burst forth at once, it might have wiped the River Rouge plant off the map.

The one man at the plant who couldn't perceive this was Ford himself. His mechanical obsession turned finally into a monomania. He came to love the sight of driving pistons and the sound of whirring wheels. He was a mechanical Captain Ahab, adrift on a steely sea, pursuing not the white whale Moby Dick but some black phantom Lizzie. Like Ahab, he could neither turn back nor ever hope to succeed in his weird quest. Ahab could tell the species of whale by the way it spouted; Ford could name the make of a machine by the way it purred. But neither special talent was of ultimate avail. At the climactic moment, Lizzie destroyed Ford with the same impersonal and evil ease with which Moby Dick sent Ahab and the *Pequod* down into the silent sea.

Henry Ford's life stretched from Gettysburg to Hiroshima. As much as any other American, he brought the two historical landmarks within one lifetime. To say this does not mean that he entered the atomic age in triumph. When he died at eighty-three he was an embittered old man. Though he worked on it for a lifetime, he failed to produce a mechanical key to happiness. As a final irony, this genius of technique died by candlelight, the electric power having failed.

No lifelong friends were at his deathbed. He had never learned that men were not machines to be tested, exploited, and junked. He had driven away his most loyal associates—even, in the end, Sorenson, Cameron, and Bennett.

Ford's boast that "his" men would never vote for a Ford union had backfired when they did so by a huge majority. His social ineptitude may well have contributed to his only son's early death. It may have been instrumental in his oldest grandson's majoring in sociology at Yale. Intricate machines purred at his bidding, but his own human thought process defied him. Never once was he humble enough to admit it to himself or his friends. The closest he came to it was when, after seeing order all about him, he remarked, "It isn't fun anymore." The statement stands as the final assessment of Henry Ford's mentality. One of his last projects was to rebuild the schoolhouse of Mary and her lamb. There was obvious irony in this, for the mind of Ford had never graduated from the little red schoolhouse.

Still Ford was a world figure, especially admired in Russia and Germany. Soviet technocrats envied his achievements. Germans lined up all night to buy securities when he built a plant in their country. Books about Ford in a dozen languages show his great influence outside America. What Ford's philosophy might come to mean, no one can say. Japan and India have added such terms as "Fordize" and "Fordism" to their vocabularies. Who knows where this might lead millions of machine-starved Asiatics?

On June 16, 2003, the Ford Motor Company celebrated its centennial. What a century it had been for Ford! Ford's great-grandson, Bill Ford Jr., is still chief executive. The company that began with ten employees in a converted wagon factory in downtown Detroit now operates in more than 200 markets on six continents. Starting with $28,000 and a few tools, it has become a $163 billion global enterprise with 350,000 employees. No one can measure its impact on popular culture.

Ford enters its second century with eighty-seven different vehicles under eight brand names. Though none may ever be as famous as "Tin Lizzie," the nickname of the Model T, later names such as Thunderbird, Mustang, Pinto, Bronco, and Explorer are part of the American lexicon.

Henry Ford has passed into the realm of legend. To different groups he represents different ideas. To the psychologists he is a mechanical Peter Pan, whose human understanding never increased. To the economist he's the sorcerer's apprentice, who discovered how to make the broom carry water, but not how to stop the broom. Philosophers think of him as Frankenstein, who became the master and the nemesis of his maker. For the citizen who dotes on mass-produced objects, Ford looms as Prometheus who brought the secret of a new mechanized fire from heaven. To Europeans he symbolizes industrial, mass-produced civilization, the full showcase and the empty mind. More than any man of his age he embodies our technological triumphs and human failures, and the still-potent dream of a democratic society of economic equality, to be achieved through mass-produced abundance. The contrast between the beauty of Ford factories and the ugliness of their workers' lot poses a crucial problem for our time: how can the human being maintain individual dignity and freedom in an industrial society?

Ford's epitaph should read: "He put us on wheels." Already we are skeptical about where those mindless wheels may take us. We recall Will Rogers' remark that America is the only country on earth where a man can ride to the poorhouse in a Ford car. Ours is a great spiritual poverty existing in material opulence. Daily we ride back and forth to earn our daily bread. When we get a few weeks of vacation from our routines, we sit back and watch the Fords go by or we get behind the wheel of our own Ford. The intuitive mechanical genius of Henry Ford has made this possible. He also brought us a message: machinery is the new Messiah. Will Henry the Mechanic prove to have been a latter-day John the Baptist?

Folk-Joke: Joe Magarac

These pretty country folk would lie . . .

William Shakespeare
As You Like It

He has faded now, but for many years, Big Joe Magarac was a star in folk-pop heaven. Trusted "authorities" shouted his praises. He had a place of honor in Carl Carmer's *America Sings,* Walter Blair's *Tall Tale America,* and B. A. Botkin's *A Treasury of American Folklore.*[1] William Grooper painted Joe bending a hunk of red-hot steel with his bare hands, and Frank Vittor proposed a towering one-hundred-foot Magarac statue for downtown Pittsburgh. Readers soon found him in such scholarly journals as the *New York Folklore Quarterly* and *Western Pennsylvania Historical Magazine,* as well as the *Saturday Evening Post* and comic books. U.S. Steel and Carnegie-Illinois Steel advertisements pictured him. Joe even got a governmental boost in a Federal Writers Project report. The Jinni of Steel has been adapted with equal success by class-conscious writers who have opposed corporations and capitalism; he is the hero, for example, of Michael Gold's "A Strange Funeral in Braddock."[2] On one point the evidence is conclusive: Joe Magarac had arrived. Oddly enough, no one seems to know quite how.

Despite individual twists, there is only one basic tale about him. Since variation of theme and event is the folk's specialty, and the publicist's main device, this is noteworthy. How would the folk in one remote area know just what people elsewhere had said? Equally disturbing is the fact that although Joe is always presented as a Hungarian, the name Magarac does not exist in the Hungarian language. Whoever picked Joe's name and nationality either didn't know better or else created a deliberate confusion.[3]

Writers have repeated, without giving any hint as to historic source, the notion that Joe's birthplace can be traced with some certainty to the J. Edgar Thompson Works, built in 1873 at Braddock,

Pennsylvania. Yet files of the Western Pennsylvania and the Pennsylvania Historical Societies, as well as those of the Pennsylvania Folklore Society, have nothing that confirms this "certainty." Nor do the newspapers, periodicals, or special collections in the region. If Joe flourished in the nineteenth century, not a shred of evidence is left to prove it.

What seems likely is that Joe Magarac was born in a magazine article done by Owen Francis in 1931; that Francis was tricked by his informants into using Magarac; that Jules Billard, Frank Vittor, and others helped spread the story; and that Joe has endured because the steel industry needs a "folk hero" of its very own.[4]

Owen Francis, the central figure in the Magarac story, was born in southwest Pennsylvania. After a brief schooling, he went to work in the Monongahela Valley steel mills, where he stayed until World War I. There he learned the jargon and methods of the steel industry. During World War I he was a doughboy with the 18th Pennsylvania Regiment. Badly gassed, he was hospitalized for a long time. During his convalescence, he read the works of Gogol, Turgenev, and Gorki. This interest may have been related to his contact with eastern Europeans in the steel mills. "Having breathed the literary air," Francis said, "I could not go back to slag and prune jack." After working as a day laborer, cantaloupe picker, and bus buy, he went to the University of California, where he got three years of education before becoming a Hollywood press agent.

When Warner Brothers bought one of his stories, Francis decided to devote all of his time to script writing. The movie did not work out. Eventually he returned to Pittsburgh, the mills, and the common people. He was through with the super-sophisticates of Hollywood. His ambition was to be back with "the Hunkies and the Polacks, who understand life." The sentimentality and condescension in the account prepares us for what happened when he turned his journalistic talent into folk hero channels.

Although Francis had no formal interest in or contact with folklore, he decided to invent a mythical steel hero. After revisiting the mills and settlements, he introduced Joe Magarac to the world in the November 1931 issue of *Scribners.* The motive seems plain enough. He wanted to sell a story.

Here we find the basic Magarac tale which has been only embellished by later writers. A man's man, he could outwork and outlift anyone in the mills. Born in an ore mountain, Joe was as strong and enduring as the steel he made. He lived at Mrs. Horkey's boarding house. As did all the other steelmen, he admired the lovely Mary, whose father (Steve Mestrovich) offered her in marriage to the winner of a weight-lifting contest. Joe won it, but allowed Mary to marry Pete Pussick, whom she loved. Later on Joe threw himself into a furnace to improve the quality of steel for a new mill. Like John Henry, he gave his all to the job.

This martyr's death has been variously interpreted. Staff writers for U.S. Steel modified it so that, when the Depression came, Joe melted himself down to make better steel for a modern mill that would produce more. Thus Joe was made to support the industry's argument that hard times are cured by more production and lower prices. The less capitalistic Federal Writers Project had Joe realize that his great strength was actually depriving others of jobs. His solution was to quit and hibernate until prosperity and employment returned. He was an NRA hero.

All this was very well, but the fact remained that the people who should have known most about Joe, and been closest to him, knew nothing. Editors of Slavic and Hungarian newspapers had no information on him. Local historians and editors of Braddock, McKeesport, and Homestead (the Magarac towns) had never heard an oral story, nor had the workers. Folklorist Hyman Richman interviewed a hundred steelmen (Swedish, Croatian, Slovenian, Polish, Russian, Hungarian, and Italian) without finding one who had heard of the steelman's idol. Nor did the dictionaries help.[5]

In one instance, Richman was able to interview forty-two informants, the entire supervisory staff of a large mill. They had from twenty to fifty years of steel experience, many having worked up through the ranks. Not a single one had ever heard of Joe Magarac.

It is, in Serbian, a word whose closest English equivalent is "jackass." Paul Blazek, Slavic publisher, said that to call a man magarac "is to lower him to dirt that is worse than after pigs pass over it." An old-time steel worker in Clairton was embarrassed when Richman used the word in front of his wife. Steve Barko, a molder, warned him that if he called someone magarac he might get his head bashed in.

What probably happened is that the Slavic steel workers decided to play a joke on the tenderfoot—in this case journalist Owen Francis. It was like telling him to go get a sky hook, or a bucket of steam. They must have been amazed to find a former mill worker so naïve as to be deceived by something they reported in jest. Albert Stolpe, veteran steel worker, told Richman, "Somebody played a hulluva big joke on Francis."6 Andrew Matta, a Braddock mill's craneman, thought they wanted to make a fool out of him. Not only Francis, but also many American folklorists and anthologists were fooled. The joke, and the hidden profanity, stuck.

With this in mind, we can, in reading Francis's other Magarac articles published in *Scribners* from 1930 to 1935, detect his gullibility, and his attitude toward his ethnic informants. He refers to the "child-like delight" of the Slavs, in much of the condescending manner that Rudyard Kipling used in describing the Africans in the days of colonialism. Frequently Francis uses the word "Hunkie," which is degrading and insulting to the Hungarians. Never penetrating the minds of the people with whom he talked, he came to collect their material, not to understand them. Such a man might well accept seriously, and put into print, what he did not recognize as a joke at his own expense.

Several years later a clever writer named Jules Billard got hold of Francis's brainchild and decided to do an article on Joe. Though he did not bother to acknowledge his debt, he showed it in his phraseology and displayed it to several million *Saturday Evening Post* readers, thus helping to increase Joe's fame. Billard professed to be an admirer of this man of steel. It was a remote sort of admiration, since he apparently didn't get closer to the steel workers than Francis's article. He merely toned down some of the lustier passages.

For example, Francis's Magarac had said, "'My name is Joe Magarac; what you tink of dat, eh?' Everybody laughed at 'dat,' for magarac in Hunkie means Jackass Donkey. Dey know dis fellow is fine fellow all right when he say his name is Joe Jackass."

Translated by Billard, the passage became: "'My name's Joe Magarac,' he said laughing. Instantly the crowd's tension snapped. Everyone knew that a fellow with a name like Joe Jackass was an all-right guy. It's a compliment to be called magarac."

Another man, himself an immigrant from southern Europe, had meanwhile seen the heroic possibilities in Joe. This was the Italian-born sculptor, Frank Vittor. It was he who headed the campaign on behalf of a monumental Magarac statue "worthy of the age of steel." To be executed by Vittor, it would be placed at the proposed Point Park where the Monongahela and Allegheny rivers merge to form the

Ohio. Thus the steelworker's idol would look out over an America which his fortitude had made possible.

In 1953 Vittor unveiled a clay model of just what he had in mind. The severest critic had to admit it was monumental. To be made of steel and bronze, it would tower one hundred feet in the air, cost three million dollars, and take five years to complete. No one has suggested that Vittor get started. But even if he ever does, the publicity and attention he had created place him among the Magarac hero makers. Like Joe himself, Frank Vittor sees things on a large scale.

The big steel companies have found Joe ideal for advertising purposes. "This Side of the Iron Curtain and Glad of It," which appeared in the July 1948 issue of *U.S. Steel News,* found in Joe Magarac the refutation of another powerful Joe in the USSR. Hardly a month goes by that Magarac doesn't break into print on the steel front. His position as the folk hero of a major industry that ought obviously to have one is taken for granted. Demand and logic have created what the folk did not. Joe has come from the top down, rather than from the bottom up. The Magarac tale has not yet entered the workers' oral tradition, even more than two decades after Francis's work, and probably never will. It is too phony. But enough elements from the older Bunyan and Crockett traditions exist to keep the general reading public interested and unaware of the hoax worked first on Francis, then on folk writers, and finally on the public itself.

ARF

There are a great many odd styles of architecture about.

John Ruskin

Architecture is the mother art. All other arts, artists, and artifacts seek some kind of shelter; we live, work, play, study, and die in buildings. Poetry hides in books, plays and ballets in theaters, paintings in museums. Buildings and bridges can't hide. They are in the public domain. Doctors bury their mistakes; architects build theirs.[1]

We identify civilizations by structures—China's Great Wall, Egypt's pyramids, Europe's cathedrals, America's skyscrapers; we link Rome with the Colosseum, Paris with the Eiffel Tower, London with Big Ben, Houston with the Astrodome. The masonry is the message.

Great architecture is irresistible. Procopius tells us that "Rude ambassadors from the Germanic north" fell unconscious with awe when ushered under the dome of Justinian's Hagia Sophia in Byzantium. Centuries later a visitor to the monastery at Cluny wrote: *"At mox surgit basilica ingens"*—surges the mighty basilica. "I sat for hours gazing at the Maison Carree like a lover at his mistress," wrote Jefferson from southern France. Anyone tired of gazing at Roman ruins can go to Versailles and send postcards to impress the folks back home.

No thought of "popular housing" in all this. No reference to the thousands of homeless who clutter the streets, alleys, and grates of America in the cold. Where *have* the "ordinary people" lived, loved, and worked over the centuries? Should not *this* be a concern of popular culture?

Nor should we limit ourselves to America. Look at any third world country today. See how little of our vaunted wealth and technology has penetrated the architecture. Neither capitalism nor socialism has solved the problem of what Walt Whitman called "the people en masse." Consider Outer Mongolia, perhaps the only modern state where horses outnumber humans. Determined to bring the nomads into cities, where they work in factories, the socialist government has created gigantic slums that have worsened, not bettered, Mongolians' lives. None of this gets into our courses on architecture, however. We prefer Buckingham Palace and Versailles.

This "leap from monument to monument" approach was how I learned architectural history, and how I taught it to my students in the 1960s. The course climax came when we reached the International Style in the twentieth century; glass, steel, and concrete came to-gether to create the New Jerusalem. All praise to three demigods most responsible: Le Corbusier, Gropius, and Mies van der Rohe. Let America learn from Europe. Less is more. Perfection achieved through geometry!

I saw none of this sterile perfection in my quiet beautiful little col-lege town. Never mind—perhaps we could take a field trip to New York or Chicago and see *the real thing*. Then came a bolt out of the blue. In 1961 a leading architectural historian, James Marston Fitch of Columbia University, broke ranks. He wondered if modern archi-tecture wasn't going up a blind alley. Perhaps our obsession with pure geometry made much modern architecture "a sort of Procrustean template which falls like a murderous cookie cutter across those liv-ing processes which do not happen to conform to its outlines."[2]

This criticism was couched in the proper academic jargon. Not so with my former student, Tom Wolfe, whose 1965 essay was titled "Las Vegas (What?) Las Vegas (Can't Hear you! Too Noisy) Las Ve-gas!!!!" Then came his thesis: "I call Las Vegas the Versailles of America, and for Specific Reasons."

This essay, and a later book called *From Bauhaus to Our House* (1999), made Wolfe a chief definer and defender of popular architec-ture. For that he paid a price. He was ridiculed and reviled by the Es-tablishment—people such as Lewis Mumford, for whom Las Vegas was a virulent example of "Roadtown," an "incoherent and purpose-less urbanoid non-entity, which dribbles over the devastated land-scape."[3] Peter Blake's *God's Own Junkyard: The Planned Deteriora-tion of America's Landscape* was published in 1964. But this elite diatribe by the managing editor of *Architectural Forum* had little in-fluence. The crucial book started by affirming Tom Wolfe and pushing the case for an indigenous style even further.

Called *Complexity and Contradiction in Architecture* (1966), its author was Robert Venturi. Two years later he and his wife, Denise Scott Brown, began teaching a seminar at Yale and in 1972 they published *Learning from Las Vegas*. For them the symbiosis of architecture and popular culture was inescapable: "I would say that we are taking a very broadly based thing, which is the popular culture . . . and we're trying to make it acceptable to an elitist subculture, namely, the architects and the incorporate and government decision makers who hire architects."[4]

We should recall the cultural setting in which Venturi and Scott Brown were working. These were the years when pop art flourished, when rock 'n' roll emerged as *the* American music, and when the proles finally had enough money to buy pizza, gamble in casinos, and build monuments to their lifestyle. American culture would never be the same again.

Alan Gowans saw the rise of pop architecture in an even wider framework. The revolt against elitism, lasting two hundred years, climaxed in the 1960s. It abandoned almost all the traditional social functions of what through 6,000 years of history has been called "art." What it does is something quite different, whether called "art" or not. If you want to find arts corresponding today in function to the "arts" of history, activities satisfying the needs of society at large for substitute imagery, illustration, beautification, persuasion, and visual metaphors of conviction, you must look to the popular or mass arts of our time.

Looking specifically at architecture, Gowans noted that the austere International Style by the 1950s was in fact, if not in official theory, accepted as a working principle in all governments. All official architecture shows it—shopping centers, malls, model towns, etc. But in the 1960s the primitivism of anarchists of the 1890s (Paul Gauguin, for example) became the fashionable mode in art schools. Primitivism in the form of brutalism, pop, and existential expression manifests it.[5]

Brutal, pop, existential? What did all this translate to out where people live, eat, and get gas?[6] After pondering the question I came up with the answer: ARF—the Architecture of Realistic Fantasy. These are structures that *look like* the activities or products associated with them, the shoe house of the Old Lady Who Lived in a Shoe, for example. Such fantasies have jumped out of nursery rhymes onto the American landscape. Next to the University of Delaware football stadium is a football-shaped diner. Down the road is a "chuck wagon," shaped like one of the old shakeguts, to be entered through a door which is a giant coffee pot. Hollywood's Brown Derby is really a large derby. The New Jersey resort area boasts of a hotel so elephantine that you can eat in the trunk. Long Island, famous for its ducks, has a duck-shaped restaurant. Nashville may not be Athens, but it has a perfect copy of the Parthenon. ARF!

Then I began searching out other ARF structures. They cropped up everywhere. Hot dog stands shaped like hot dogs, Chinese restaurants like pagodas, Indian souvenir shops like tepees. Bizarre examples cropped up, especially in California: an ice cream shop built in the shape of an ice cream freezer, a flower shop in the shape of a rose. Then I realized that ARF could influence a whole chain of buildings—the Long John Silver's chain, for example, designed to look (more or less) like a ship, with port and starboard running lights. How about theme parks, in which you could go "back to Europe" such as Busch Gardens near Williamsburg, or nineteenth-century America in Disney World?

City planners think big; the city-as-carnival idea is spreading across the land. If Las Vegas is the prototype, consider Seattle's Space Needle, Houston's Astrodome, St. Louis' Arch, and Atlanta's Regency Complex.

And what are we to make of the filling stations and motels that are central to Everyman's life in a mobile culture? The question was raised in 1974 by Bruce A. Lohof, when he published "The Service Station in America: The Evolution of a Vernacular Form" in *Industrial Archeology.* The proper study of modern and mobile societies should include not only the stately skyscraper, Lohof argued, but also the lowly filling station. "These stations are symbolic of a contemporary, motorized people, for in a literal sense they pump the lifeblood of their mobile society." He in turn made use of John A. Kouwenhoven's 1948 study *Made in America,* which defines vernacular as the art of a people living under democratic institutions in an expanding machine economy. ARF is a branch of the vernacular.

Meanwhile Arthur Berger was beginning to study motels for what they tell us not only about popular architecture but sex, ethics, gender, cars, and digestion. Today's family, he noted, tends to be a group of motel users, clustering around a private cafeteria. Mom is the short order cook, Pop the motel operator. When the kids marry, they set up their own little motels and even ask Mom and Pop over for a quick meal. Do we fantasize motels as homes away from home? As hide-aways from reality?

Next to the motel is the fast-food restaurant. What do the architecture and interiors of these omnipresent structures tell us about us? Philip Langdon explores that question in his 1986 book called *Orange Roofs, Golden Arches*. The more we explore the Strip outside town, the more we discover about the Architecture of Realistic Fantasy. Nestled beneath their transcendent arches and American flag, enclosing immaculate rest rooms and floors, radiating health and warmth, McDonald's combines many of the ritual and symbolic aspects of religion. Could it be that this is the poor man's cathedral, and it is here that he eats the food that provides common expectations and experience?[7]

ARF raises many other questions. Are our vernacular structures variants of buildings found in other cultures, or are they uniquely American? Who actually makes major design decisions for ARF, and on what information? What moved us from Howard Johnson colonial to McDonald's modern? How do stage, film, and TV sets influence architecture on Main Street? Can ARF be applied to automobiles, as well as static structures? Why do architectural schools and teachers largely ignore ARF, and cling tenaciously to Bauhaus? How do land developers, contractors, and realtors collaborate in design matters? How many decisions are local, how many are made in distant design factories? What is the model on which popular artifacts and prototypes are built? Can we conjure up a structural model which reveals the essential unity of a building, area, or procedure?[8]

Buildings are also like falling stars, easy to see, hard to explain, quick to vanish. Instead of deploring "the malling of America," we should try to explain and understand it, then perhaps to design something better.

"Architecture goes beyond utilitarian needs," Le Corbusier pointed out, "to the spirit of order, the unity of intention." Since its primary goal is to meet human needs, architecture is a social art. Indigenous buildings speak the vernacular of the people. The untutored and intuitive work of anonymous architects, Sibyl Moholy-Nagy has argued, fulfills an ideal standard. Their work is preserved and popular because of its adequacy beyond the life of the builder.[9]

Herbert Read carries this line of thought farther in *The Grass Roots of Art* (1961) and insists that architecture is always best when it is "people's art." He quotes from E. G. Coulton's *Life in the Middle Ages:* "Everybody watched the builders at work; everybody was interested in them. . . . There was not only the higher status of the workmen, but more unity of spirit between them and the public."

One senses some such "unity of spirit" in American make-believe settings: Dinosaur Park, Disneyland, or Bonanzaland. On a less popular (but no less fake) level are "historic" restorations such as Williamsburg, clean, Cartesian, sterile. I have an old friend who lived in the real Williamsburg before Mr. Rockefeller took over. "I liked it a lot better then," my friend says, "when it was dirty!" But the restoration that ARFs for us today must be clean like the floors and laundry on the TV commercials. We live, like our Victorian ancestors, on Great Expectations.

That is why a general alarm was spread when Nashville's Grand Ole Opry House, representing a far different America from that of Williamsburg, was threatened with destruction in 1973. Constructed as the Union Gospel Tabernacle in 1892 with curved wooden pews and a "Confederate Galley," it has become the mecca of country western music. Hearing that some of the bricks of the bulldozed building would be used to build The Little Church of Opryland, and that Tennessee Ernie Ford would sing the first hymn, Ada Louise Huxtable wrote, "In today's world only the phony is real. Gentlemen, for shame. You at least should be on the side of the angels."[10]

Whether you travel to Williamsburg, Nashville, or the hundreds of other recreations and restorations that dot the American landscape, you will pass giant billboards which are not only part of our popular architecture, but our site planning. Many old barns have ads painted on their sides. Madison Avenue murals and folk-fake symbiosis! Architecture of Realistic Fantasy!

Can we find horizontal parallels across history that will replace the cyclical or line-of-progress patterns? Does popular architecture hold keys to doors we have not yet opened? Can we come to terms with those "dream palaces" built to show movies in the days before TV? "We visualize and dream," a theater architect wrote, "a magnificent amphitheater under a glorious sky in an Italian garden, in a Persian court, in a Spanish patio, or in a mystic Egyptian templeyard, all canopied by a soft moonlit sky."[11] ARF! ARF!

Television married evangelism and begat televangelism. One showplace is Robert Schuller's Crystal Cathedral, only a stone's throw from Disneyland in California. Here Schuller delivers his "I Am the American Flag" sermon, which won a Principal Award from the Freedom Foundation. Jimmy Swaggart, another superstar, has a favorite quote: "If Jesus were on Earth today, He would be on TV." (*Source:* Schuller Ministries photo.)

Thunder from the Pulpit

He maketh his ministers a flaming fire.

Book of Common Prayer

Religion has always powered America.[1] Our Pilgrim fathers came as religious dissenters. Religion was a major factor for generations and still is. Frontier revivals energized the West. Later the special fervor of revivalism would move back East to the urban centers. Like Proteus, popular religion takes many shapes and forms. It sweeps ahead like a hurricane. In its path come hysteria, confessing, shrieking, speaking in tongues, barking, jerking. Who can explain it all?

To the Four Gospels a fifth was added after the Civil War: the Gospel of Success. Best sellers had titles such as *Success in Business* (1867), *The Secret of Success* (1873), *Successful Folk* (1878), and *The Law of Success* (1883). Popular religion and economics merged. Preachers perfected the formula that crowd-rousers understand. Follow three things: the flag, the crowd, and the money.

For centuries success and salvation have been major American themes, merging politics, religion, and economics, ever since the Puritans landed on our shores. America was the safety valve, missionary field, haven for every missionary group. It still is. Multitudes fled to our shores, dissidents from Protestantism, Catholicism, Judaism; the Shakers and Quakers; Pentecostals and Penitentials; Baptists and Anabaptists; Mennonites and Hutterites; all sorts of Anglicans, "low and lazy, middle and hazy, high and crazy." To the far left were the Ranters, Ravers, and Diggers. Turn on your radio or television; the ranting and raving continues.

Deep in our democratic faith is a strong supernaturalism. We are bathed in the blood of the lamb. We remember General George Washington not only on his horse, but on his knees at Valley Forge, praying for God's support.

Every president since Washington has honored deity related to law and order; but popular religion has favored a God of thunder. Ours is officially a civil religion, as set forth by Jean Jacques Rousseau in *The Social Contract* (1762). But, as we shall see, popular religion has soared with the American eagle—an eagle much given to screaming.

Leading the pack were Willie and Billy—William Jennings Bryan and Billy Sunday, who boasted that he knew no more about theology than a jackrabbit knew about Ping-Pong. But Lord, how the preacher and the rabbit could jump! The people loved it. Later on they would love a host of spellbinders, including Aimee Semple McPherson, Father Divine, Father Coughlin, Oral Roberts, Rex Humbard, Fulton J. Sheen, and Billy Graham.

Our Great Awakenings occur about every fifty years. The first came in the middle of the eighteenth century, led by George Whitefield and Jonathan Edwards, famous for delivering the most famous American sermon: "Sinners in the Hands of an Angry God." You must be reborn if you would be saved. The sermon was not only a work of art, it was also the foundation on which American evangelism is based.

The second Great Awakening, sweeping over the frontier, was based on camp meetings and circuit riders. A new gospel hymn filled the land:

> We'll build our camp on this rough ground
> And give old Satan another round.

During the first Great Awakening God was "prayed down"; now he was "worked up." Preachers had to sew the seeds and reap the harvest overnight. That called for hell-raising and heart-breaking.

Up sprang Francis Asbury, the "Father of American Methodism." In the 1770s Methodists had a few hundred converts and four preachers. When Asbury died in 1816, they had 214,000 members and over 2,000 ministers. Many of them began their services by having congregations stand and sing:

> The world, John Calvin, and Tom Paine
> May hate the Methodists in vain.
> I know the Lord will them increase
> And fill the world with Methodists.

Time flies, conditions change. As the frontier moved west, religious fervor moved east. The third Awakening occurred in urban industrial America. Charles Finney led the way, adapting the old frontier tactics to the new cities. Both frontiers were vibrant and vital. The year 1856, the *annus mirabilis* of revivalism, transferred the "rural love feast" to the cities, where "revival fever took hold." The YMCA and the U.S. Christian Commission flourished. Billy Sunday gave it mass appeal and motivation, and the "penny press" (cheap newspapers for all) carried the message. There were powerful new ways to proclaim God's wonders.

Fundamentalism powered the fourth Great Awakening in the early twentieth century; William Jennings Bryan, the Great Commoner, led the way. He, Father Divine, and Aimee Semple McPherson guided millions through the Great Depression and World War II.

The fifth Great Awakening came with the Age of Ike, a rich triumphant America, and the emerging of new phenomenon: the Electronic Church, and televangelism. From Puritans in somber black to TV "stars" in living color and new thunder from the Pulpit. The 1970s and 1980s were glory days, when electronic religion assumed prominence and power beyond anyone's expectation.

How did this happen? Who were the people that guided it, and why did some of them fall as quickly as they had risen? Will it regain power in the twenty-first century? What does it tell us about the power of popular culture? Will there be another Great Awakening?

Old style elite theologians and church leaders watch their membership and influence decline, and condemn the new televangelists and evangelists who prosper like the green bay tree. The new Age of Circuitry has altered popular religion, like everything else.

In July 1955 Thomas J. J. Altizer was reading a passage by Friedrich Nietsche. "God is dead," jumped out from the page. "I knew that God was dead," Altizer later wrote, and a whole school followed built on that alarming conclusion. A new thrust in theology turned from dreams of a New Jerusalem to the Secular City.

Hold this thesis up to the twentieth century. Many believe the City of God has become the Secular City. Instead of depending on the Almighty, we depend on the technique to solve problems. Forget ultimate goals; go for the practical and expedient. Yet millions will not abandon the City of God and the Heavenly City. While liberal religions decline, fundamentalism and the religious right flourish throughout the world. Christianity, Judaism, Islam, and Buddhism attest to this. Powerful movements are affecting not only religion, but politics and public opinion. In Israel, India, Tibet, and the United States—to name four obvious examples—power centers must deal with fundamentalism to survive. In America the "Moral Majority" reversed the sixty-year hold of New Deal Democratic politics and issued in a new era.

Tracing the tide of fundamentalism on the world scene is a vast task still to be done. For our mosaic, I shall attempt to deal only with my own country, summarizing my longer account, *Great Awakenings: Popular Religion and Popular Culture.* Twentieth-century "ordinary Americans" struggle with questions that bothered our seventeenth-century Pilgrims and Dissenters. What happens to sinners in the hands of an angry God?

The cult issue had rested on the back-burner for most Americans in the 1990s. The economy was booming, the stock market zooming, and there were a million new millionaires. Cult Christians, after all, were mainly ignorant hillbillies who liked to handle snakes. All that changed in February 1993. A group of cultists called Branch Davidians, led by self-proclaimed prophet David Koresh, were holed up at Waco, Texas. Their children inside were being coerced and perhaps abused.

Call the FBI.·So down to Texas went a strong-armed contingent to clear up the problem. On February 28, 1993, four agents from the Bureau of Alcohol, Tobacco, and Firearms (ATF) stood at the locked door and demanded entry. Minutes later all four were shot dead. Now America *was* tuning in on Waco. A media-saturated fifty-one-day siege followed. Could it be that the cultists had stood off the FBI? Attack!

So they did on April 19, 1993. Something went wrong. When the episode ended, seventy-seven people, including women and children, had died during the fire. Years later, an embarrassed and frustrated Congress was asking, was this the slaughter of the Innocents? How can we cope with tongues of fire? With strange words and no words?

We must beware of concluding or condemning too soon. Is popular religion mass religion, controlled by mass media, or religion "of the people," controlled by the people themselves? When we measure growth and impact by ratings, receipts, and polls, are we dealing with substance or surface scratchings? Which voices will endure in a society where everyone is famous for fifteen minutes?

Will the wall which Jefferson helped build between church and state crumble? Answers to these questions will help shape America in the twenty-first century.

Robert Schuller's Crystal Cathedral in Garden Grove, California (*Source:* Schuller Ministries).

LOOKING AHEAD

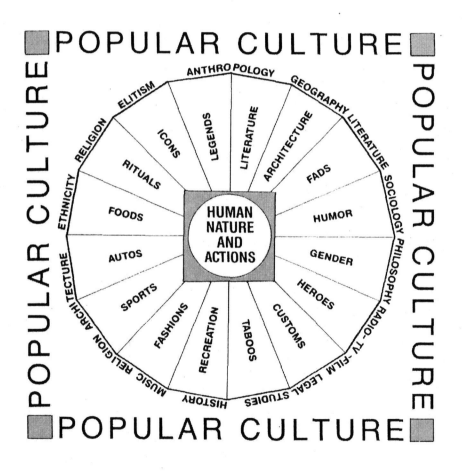

Living with Machines

You're not a man; you're a machine.

George Bernard Shaw

There is something fascinating about machines. By the thud of machinery and the shrill steam whistle, Walt Whitman wrote, we are undismayed. Over a century later Andy Warhol went a step farther. He said he wanted to be a machine.

Some question if we can live with them, but who can doubt that, in the twenty-first century, we can't live without them? They are omnipresent in our world, our lives, our homes, even in our bodies. We live in the Age of Machinery, in every inward and outward sense of the word.[1]

Every new machine replaces not only older machines, but in some instances a "way of life." A famous early example is the spinning jenny, which changed lives and communities in England. In the early Industrial Revolution, the legendary Ned Ludd went so far as to lead a revolt to destroy the machines, to preserve the handicrafts and cottage industries and restore the English countryside to its ancient ways. Of course, the machines won.

Many years later another Englishman, Aldous Huxley, warned against a machine-dominated soulless *Brave New World*. Published in 1932, the book is still popular and disturbing. Many leading thinkers, such as Lewis Mumford, Christopher Lasch, and Neil Postman have affirmed Huxley's fears. Of special importance is a pivotal work by Neil Postman, *Technopoly: The Surrender of Culture to Technology* (1993).

The Electronic Revolution, starting with Edison's lightbulb and moving quickly to Bill Gates's computers and instantaneous worldwide communication, wiped out in one lifetime what had been aspects of Western culture for centuries.

In that same lifetime we have witnessed the rise of utopian machinery, better known as science fiction. Buck Rogers and his helmeted cowboys went into space, and real human beings soon followed. We walked on the moon. Suddenly we find ourselves upgraded, downloaded, networked, digitized, living in cyburbia, thrust out into cyberspace.

Not without a price. Unexpected new problems, some defying solution, pop up. Hackers break into secret files, blackouts turn off the switches, computers crash. We give to our trendy machines a new label: transitional technology. CDs, laptops, Unix tapes, diskettes, Betamax, eight-track tapes rush in and fade out, sometimes in a single season. One of the computer pioneers, Ted Nelson, summed up our twenty-first-century dilemma: "The so-called Information Age may be the Age of Information Lost."[2]

But have we lost far more than information? What used to be called "morality," on which our "American way of life" rested, seems to be melting away. Violence dominates our news, television, and popular films. Families dissolve, children are kidnapped or even murdered. We put "In God We Trust" on our currency, being "one nation, under God" in the Pledge of Allegiance. Efforts to remove "God" from both surfaced in 2002. Whom *do* we trust?

It took centuries to move from the natural to the mechanical environment, but only a few years to move to the electrical. The magnoelectric revolution occurred in the twinkling of an eye and blinded many who witnessed it. Pressing questions beg for answers. Will the new cyberculture, wiring the world into a Global Village, make much of our earlier lives and beliefs obsolete? Will we leave what we called the "modern world" behind and rush headlong into postmodernism? Have we done so already? For a vivid account of what it might be like in that world, see Peter Sacks, *Generation X Goes to College* (1996). And take a good look at the world around you.

Instead of machines, much of the talk will be about technology. That is what runs, and indeed controls, our world. Is it even a new religion? But do we want to worship our machines, by any name? Like it or not, by night and by day, we have come to know how much depends on that technology and know-how. It is a series of events and techniques for duplication and multiplying materials, sounds, and images.

Nothing has been duplicated and multiplied more than monsters, a point of beginning for our chapter on machine culture. The offspring of Dracula, that blood-sucking vampire, have become legion; so have the Wolfmen and Hulks. King Kong haunts our national memory; he seems by now to *own* the top of the Empire State Building. Mary Shelley's vivid portrayal of Frankenstein, the monster who took over, has haunted us for over a century. William Morris's *News from Nowhere* (1890) described irate workers smashing machines to return to the "good old days" of handicraft—a theme continued in Samuel Butler's *Erewhon* (1872), W. H. Hudson's *A Crystal Age* (1917), and E. M. Forster's short story "The Machine Stops" (1909), all of which portray technology as a habit-forming drug.

No longer is it fashionable to shudder at every scientific advance, and note that we're just one step closer to George Orwell's *1984* or Aldous Huxley's *Brave New World* where sex, sports, and soma prevail.

> Orgy-porgy, Ford and fun
> Kiss the girls and make them One.
> Boys at one with girls at peace;
> Orgy-porgy gives release.

But do we want the kind of release found by a nine-year-old boy named Joey, who wanted to be run by machines? He put together an elaborate life support system composed of radio tubes, lightbulbs, and a breathing machine. Joey ran imaginary wires from a wall-socket to his stomach, so he could digest his food. His bed was rigged with batteries, a loudspeaker, and monitors to keep him alive while he slept. Joey was living out his autistic fantasies in a machine culture.

Such alarming episodes have caused some to reject science and its enslavement, to hark back to Rousseau, and the Noble Savage. But if these romantics break a leg, will they forgo the hospital? Or take an antibiotic for an infection? Or go west by oxcart instead of jet plane?

Technology is a branch of moral philosophy as well as science. Great thinkers of the past have not sought technological humans, but humanized technology. "If we continue to place God in the machine," Roger Burlingame warns in *Backgrounds of Power: The Human Story of Mass Production* (1949), we shall be at the mercy not of the machine, but of its high priests who know better and are thus in a position to exploit our ignorance."

People want to turn to something, or somebody, who has the answers. For traditional man, that somebody was God, for modern man, science. He puts self above neighbor, state above church, and science above all.

Intrigued by new gadgets, drugs, metals, and shortcuts that scientism (false science) provides, the public values applied sciences over the basic sciences. Results are appalling. Landscapes are strewn with outmoded eggbeaters, last year's neon signs, crinkled automobile fenders, empty beer cans.

Getting and spending, we lay waste not only our powers but also our happiness. *Things* are substitutes for pleasure. For Keats a thing of beauty is a joy forever; we have shortened the line: a thing is a joy. If it threatens to last forever, we junk it for a shinier model.

Science, having proclaimed God's power in the hands of humble men, can become itself a false messiah in the hands of greedy ones. It's the old Baal business all over again, chrome-plated, jet-powered, gimmick-laden. The new commandment is plain: love things and use people. The reverse proposition, to love people and use things, was outmoded.

Scientism has opened a gap between science and ethics. Men have produced hydrogen bombs while admitting that no moral society would sanction their use. In their efforts to preserve humanity, they find themselves becoming less than human. Bombs keep coming. We have a job to do, their manufacturers say, as if they do not also have a responsibility.

Scientism has leaped out of the laboratory into classrooms, living rooms, and churches. It has been as attractive to communists as it is to capitalists. An unending international atomic squabble is inevitable. Both blocs are primarily devoted to *things* to solve their problems and impress their allies.

Scientists of our times have warned us. The picture of Albert Einstein in tears before a Congressional committee and of Robert Oppenheimer in disgrace may seem to future historians two of the most significant figures of the twentieth century. "What have you done to my people?"

Some of our most creative technological minds are striving to bring variety to chromobile tail fins, and to make the refrigerator light come on half a second quicker when we open the door. Is it for this that Galileo and Descartes and Einstein labored? Science hesitates to answer. Scientism does not. The answer is on the wall, in the form of a rising line on a graph: one answer, at least, though the final answer may be quite different, in a place where new models are not nearly so relevant as old mistakes.

Meanwhile, new models flood the market. We buy the car of tomorrow, only to find out tomorrow that it is the car of yesterday. We dispatch it to the junk heap, and return like Sisyphus to the foot of the hill (or loan office) to begin all over again. The mad dash goes on forever, and our chase has a beast in view.

Man against the beast—not only the beast within him, but the beast riding in the back of the police car, trained, ready to leap and (if necessary) to kill. Will today's police dog become as critical as the sheriff's horse a century ago in maintaining human decency? Or will zebras learn to respect one another's stripes, be they black or white? Can we rebuild and retool in technopolis?

This question need not dismay us, any more than retooling for a new automotive age need terrify Detroit technicians. No one is asking or expecting us to give up our archetypes, our way of life. When car manufacturers move engines to the rear, the cars still go forward. We may abandon the killer tactics of the tiger and still survive and prosper in the complex new jungle.

America's technician mentality shapes not only her assembly lines and vending machines, but all aspects of her culture. Apply that mentality to painting and sculpture, and pop art emerges. Technology creates its own sense of time: dynamism is built into the model. Change becomes an annual ritual, a confirmed habit, an absolute. Innovation is the rule. Styles hardly come in before they move out. Color me culture lag.

The technician mentality abandons traditional notions of myth. Transcendence can't be put on a blueprint or flowchart. The power that technology puts into man's hand is matched only by lack of understanding as to its meaning. Just as it is risky to repair electrical fixtures without understanding electricity, so is it folly to play with new social forces without understanding their meaning. Might this help explain our lack of viable mythology? And the turn to the occult, irrational, and emotional by some children of technocrats?

The question points out a baffling paradox. There are others: we live at a moment of economic dilemma and political pessimism. Automation, computerization, nuclear fission, and rocketry should have made us free but have only tyrannized us to a degree unparalleled in history. Our ancestors thought we would by now have built the brave new world. Instead, we seem bent on blowing it up. Countries such as Iraq and North Korea, determined to become nuclear, drove the world to the brink of war in 2003.

Whatever its faults, technology does change and adapt; in fact, it enforces obsolescence on the very machines which pour out of the cornucopia. The operative words are *mass production* and *assembly line.* In the popular imagination, one man gave us these things: Henry Ford.

Ford's idol, a young inventor named Thomas Edison, decided in 1875 to invent a machine that talked. He wanted a unit to record messages, perhaps even rebroadcast them. He lit on the idea of using a metal cylinder with spiral grooves, and a piece of tinfoil to wrap around the cylinder. After many tries, it worked. Edison himself knew this first. His own hesitant first five words ever recorded were "Mary had a little lamb."[3]

By 1878 he had formed the Edison Speaking Phonograph company, and written a list of ten ways in which he predicted his toy might benefit humanity:

1. Letter writing and all kinds of dictation without the aid of stenographer
2. Phonographic books, which will speak to blind people without effort on their part
3. The teaching of elocution
4. Reproduction of music
5. The "Family Record"; a register of sayings, reminiscences, etc., by members of a family in their own voices, and of the last words of dying persons
6. Music-boxes and toys
7. Clocks that should announce in articulate speech the time for going home, going to meals, etc.
8. The preservation of languages by exact reproduction of the manner of pronouncing
9. Education purposes, such as preserving the explanations made by a teacher, so that the pupil can refer to them at any moment, and spelling or other lessons placed upon the phonograph for convenience in committing to memory
10. Connection with the telephone, so as to make that instrument an auxiliary in the transmission of permanent and invaluable records instead of being the recipient of momentary and fleeting communication

By the end of World War I there were over 200 phonograph manufacturers; the list Edison had made was being turned into reality. A new music form called jazz made popular music a gold mine; over a hundred million records were sold in 1927. But the *real* boom lay ahead, when rock and roll (a blend of country music with rhythm and blues) surfaced in the 1950s. The one person who is often cited as "father of the form" is Cleveland deejay Alan Freed, who coined the term rock and roll, joined the staff of WINS in New York, and introduced Bill Haley's "Rock Around the Clock," the first rock and roll single to reach the top of the charts. Meanwhile a shrewd promoter, Sam Phillips, drove all over the South looking for "a white boy who can sing colored . . ." He found him in 1954, and recorded the performer's first hit. His name was Elvis Presley.[4]

The growth of other new electric media was just as spectacular and revolutionary. While Edison was improving his phonograph, others were developing the radio. Young David Sarnoff decided in 1946 to make radio a household utility in the same sense as the piano or phonograph, and the corporation he headed (RCA) did just that. Wireless transmission of images sped forth the growth of television. A landmark was President Franklin D. Roosevelt's televised opening of the 1939 World's Fair. The growth after World War II was incredible. Operating sets jumped from 7,000 in 1946 to around 172,000 in 1948, and over 5,000,000 in 1950. Even theologians were awed. "Radio is like the Old Testament," Bishop Fulton Sheen said, "hearing wisdom without seeing. Television is like the New Testament—wisdom becomes flesh and dwells among us."

So, by then, did cyberculture, which has to do with controlling or monitoring machines—especially computers. Information is gathered at different points along the river of communication, stored at a central pilot post or machine, then distributed to different parts of the system on request.

In cyberculture, the machine begins to take over from humans; to know more, remember more, react more quickly than can any human being. Already computers check our taxes, reserve our tickets and hotel rooms, handle our subscriptions, do our banking.

Putting it in this terminology helps show the link with the past: with the Luddites, who went around destroying the new spinning jennies because they were replacing spinning wheels in nineteenth-century Britain. What will happen when . . .

The chief argument of the antitechnologists is pro-Frankenstein. Technology is a beast, created by humans but now beyond human control: it will destroy us. Other leading antiarguments are as follows:

1. Those who work with machines become machines.
2. Not the workers, but the technocrats, benefit from this. Recall Charlie Chaplin in *Modern Times,* the prototypical image.
3. Technology forces us to consume things we neither need nor want.
4. Technology is the enemy of all things natural; God made the country; humans made the cities. The updated name of this old argument is, of course, ecology.
5. Using mass communication and mass advertising technology creates mass humanity, glued to the boob tube, enslaved by a slavish mentality.

A steady stream of authors have deplored the tendency to turn men and women into machines. Among the novelists, John Dos Passos is especially persuasive in his novel *Three Soldiers* (1921). Militarism with its ever-improving technology, he believes, ends up as a form of mutilation. Living human beings become mere pieces of machinery, to be killed by machine guns, tanks, and bombs. Dos Passos centers on World War I, but all later conflicts confirm his fears.[5]

Lewis Mumford elaborates on the theme. An army of machines is the central force in Western civilization. Consider warfare. Not only have machines killed millions; they have also depleted the countryside, destroyed the forest and whole species, damaged the ozone layer, and brought into existence irredeemable misery. Now human beings can do all this without leaving their home or office, thanks to the wired world of computers.

What is our only hope? To reassert ourselves and take back the power we once had? The central fact of existence is not machinery but life itself.

More machine worries emerged in 2003 when the Silicon Valley Toxics Coalition issued a warning about e-waste. Now e-mail isn't alone as a problem. U.S. companies release large amounts of lead, polyvinyl chloride, and other hazardous materials used in computer manufacturing. IBM computers also contain brominated flame retardants, suspected of blocking hormones and impairing biological processes. The health and environment impact of all this is yet to be determined.

The National Safety Council estimates the United States will be awash in 500 million defunct computers and monitors by 2007. Less than 10 percent will be refurbished or recycled. How's that for creating massive waste and new unsuspected dangers?

The European Union has placed the burden of recycling on manufacturers; Japan passed a 2001 law requiring manufacturers to recycle. One thinks again of the Sorcerer's Apprentice. We can outproduce what we can consume or dispose of. Our landscape looks more and more like a junkyard, and our clinics puzzle over mysterious new ailments coming in from Cyberland. Can we turn the tide? Alas, the Sorcerer's Apprentice couldn't. Beware.

In a single generation computers have become the motor that drives much of our lives. They win wars, send spam, and validate credit cards. When they are "down," we are out of luck—and crucial information. And when one model dies, we rush to buy an ever newer, faster, slimmer computer.

That frightening word pops up: *obsolete*. We burn old tires, hide atomic wastes in caves, build fences around acres of junked cars. What happens to junked computers?

Old computers don't just die or fade away; they become a great expense, logistical nightmare, and ethical problem that can only grow worse in the years ahead. Forty million U.S. computers were retired in 2002; the projected number for 2004 is 300 million.

Georgetown University spent $50,000 last year to recycle broken computers; the University of Minnesota spent $100,000, and will spend $150,000 in 2003. Electronic waste may well be the next big environmental issue. Six states have already proposed or passed laws banning the disposal of electronic waste. Computers involve many poisons, such as mercury, cadmium, and lead.

Tons of junked computers end up in Guiyu, China, where, for about $1.50 a day, villagers disembowel them. The Basel Action Network sent a crew to China to check the results. They found that rashes, respiratory illnesses, and miscarriages are common. Lead level in the groundwater is 2,400 times higher than what the World Health Organization calls acceptable. Filmmakers checked ownership labels of junk computers in Guiyu. Some belonged to the U.S.

Defense Intelligence Agency. Forget the sleazy breezy computer ads about our new utopia. Here is technology's trail of death.

Years ago Henry Adams warned against depending on new seductive technology to control our lives and nation. He likened it to a high-voltage electric wire that we grabbed hold of, expecting to move faster, quicker. Then we found out we couldn't let go. Instead of being happy technocrats, we were electrocuted.

Wanted: A New Mythology

> Not only is God dead, but try getting a plumber on the week-
> end.
>
> Woody Allen

Woody may be right about the plumber, but he's wrong about God. He or She might be having trouble in certain churches or seminars, but not on the Internet, sports stadiums, and TV screen.

For centuries God talk was oral, and mythology was filled with creation, God tales, tall tales, and nature. Folklore was the way of transmittal. Truth was the spoken word.

Johann Gutenberg and his printing press dealt the oral tradition le- thal blows; though as we move into a post-Gutenberg world where truth is *heard* more often than *read,* may we see mythic resurrections? Since 1600 scientific scribbling humans have moved sacral religion to the fringe of life. Thermostats and air conditioners have regulated nature; dams have tamed rivers. With the enlightenment, fears of dog- mas, weather, and tempests tended to disappear. Nineteenth-century mythology came to deal more with economics, politics, and society. Depressions, treaty breakers, and empire makers created in the West- ern world a full-blown Victorian mythology. Some of it lingered on with Winston Churchill and the World Wars.

Those who made and admired that credo prospered. Britannia *did* rule the waves; the white man shouldered his burden; the sun never set on British soil. Shielded by England's navy, the United States be- came a world power, and (after a bully little war) got an empire on her own. Three generations later the British Empire had faded; the Victo- rian mythology lingers on. It is, in fact, at the center of much contem- porary American thinking. Today we and not the British are Eminent Victorians.

Though seldom read or studied in formal fashion, Tennyson, Browning, and Kipling are still very much with us. From the slums of Harlem to the rice fields of Vietnam we expect "to strive, to seek, to find, but to yield." Of course the Peace Corps is idealistic, but if one's reach doesn't exceed our grasp, what's a heaven for? How can we turn over vast tracts of Africa or Caribbean Islands to the natives, half devil and half child? Technicians girdle the globe to "Ring out the old, ring in the new," while public health experts "Ring out old shapes of foul disease." When he wants to reason with the world, does not President Bush paraphrase Tennyson:

> Ring out a slowly dying cause,
> And ancient forms of party strife;
> Ring in the nobler modes of life,
> With sweeter manners, purer laws.

In at least seven major ways Mr. America tends to reflect Victorian mythology:

1. He believes in progress, and expects his latest gadgets to move onward and upward.
2. He cherishes the self-help cult and likes men who "come up from the ranks."
3. He is self-critical, yet always reaffirming in the end.
4. He believes in world order as laid out on his own blueprint.
5. He always glories in means (i.e., techniques) more than ends.
6. He is unabashedly materialistic.
7. He is jingoistic.

Look at American life today and you will have no trouble recognizing the expressions for basic tenets listed here. Mass circulation magazines, newspapers, speeches, manifestos, Chamber of Commerce bulletins, labor union brochures, and government press releases reaffirm the old, remarkably durable mythology. An unmatched system of productivity, victory in two world wars, freedom from devastation during those wars, technical ingenuity, and vigorous leadership have allowed America to keep certain nineteenth-century beliefs intact long after Europe (which originated them) abandoned them. Now history is catching up with us too. Even though the marines have landed, the situation is *not* well in hand. Our generation must finally meet the world's terms. We must remythologize.

The building blocks of mythology are archetypes—inherited ideas (or modes of thought) derived from long experience. Present in the unconscious thought of the individual, they control ways of perceiving reality. Western Man is himself a sort of archetype: mobile, optimistic, materialistic, nonaesthetic, and amoral. His trademark is energy; his goal is power. We confront him in literature as Faust, Ahab, or Frank Cowperwood, and in pop culture as the Lone Ranger, Pac Man, or Agent 007. As comic strip writers and TV producers know, archetypes can quickly degenerate into stereotypes. Then they lose *manna,* and become targets for parody. Going back to the original inspiration, a new archetypal model must be invented: sometimes even a new archetype. Self-renewal is not only the key to variety; it is the road to survival. It is more important to construct archetypes than atomic weapons.

The prospect need not terrify us, any more than retooling for a new day terrifies Detroit automobile manufacturers. No one is asking or expecting us to *give up* our mythology, our "way of life." When manufacturers moved engines to the rear, we still expected cars to go forward. We retool not to destroy, but to survive.

American culture's two major twentieth-century remythologizers, Franklin D. Roosevelt and John F. Kennedy, are dead; others will follow and supplement them. The book of one of our ablest political leaders, Senator J. William Fulbright, had the significant title *Old Myths and New Realities* (1964). He points out that the fairly short list of "self-evident truths" of the Founding Fathers had been vastly expanded by 1965 to include, for example, the Cold War. Just as the President resides in Washington and the Pope in Rome, so the Devil resided immutably in Moscow. Confronted with a complex and fluid world situation, we are reluctant to adapt ourselves to it. Instead we cling to old myths in the face of new realities and seek to escape the contradictions by relegating an increasing number of ideas and viewpoints to a growing category of "unthinkable thoughts." When thoughts are unthinkable, culture cancer is incurable. We can, of course, cling to and even perish with the old myths, or we can remythologize and live.

With books such as Fulbright's available to analyze old stereotypes about Communism, foreign policy, and the Cold War, it is hardly necessary to spell out the mythic overtones in Cuba, Vietnam, and Iraq in detail. The important thing is to define remythologizing, in this context, as deliberately *thinking* "unthinkable thoughts," exploring all the options that confront us in a fluid world scene, overcoming our susceptibility to shock and panic every time something we don't like occurs. Politicians who advocate remythologizing risk censure or defeat. Yet as Edmund Burke pointed out, the real duty of a politician is not to do everything his constituents prefer, but to give them the benefit of, and be held accountable for, the exercise of his own judgment. To remythologize is to mold, not to follow, public opinion.

If only the brave dare attack our solidified myths in politics, would any but the foolhardy try the same thing in science? If modern man puts the state above the church, he puts science above them both. For many who are not scientifically trained, science is a sacred cow; today's father-confessors wear white, not black. When the white-robed scientists look up from their cyclotrons and speak, they may not be understood but at least they will be believed. The doctor is the master of our mythological realm. Like the ancient Wise One, the doctor can kill dragons (known as germs) and give access to magic potions (with unpronounceable Greek names).

We must remythologize: set aside old platitudes, challenge the new technology, probe dark areas where probing hurts, raise questions we would rather push aside. Discontent is the yeast that makes crucial change rise. Myths are what make it believable and acceptable.

Helping Humpty-Dumpty

> O! What a fall was there!
>
> William Shakespeare
> *King Lear*

We think in millions, billions, and trillions: million-dollar lawyers, million-man marches, million-member credit cards, billion-dollar mergers, trillion-dollar budgets, tax cuts, and debts. The number that pales into insignificance is *one;* in what we long and proudly saw as "*one* nation, indivisible." Now one critical word eludes us: *unity.* The new buzzword, diversity, has come to mean disunity. Things continue to fall apart. Will we end up like poor Humpty Dumpty?

> All the king's horses and all the king's men
> Couldn't put Humpty together again.

That might depend on our coming to realize we can't mend Humpty. One is the most important number in the world and unity our most precious possession. One stands for a single unique entity. Every leaf, every grain of sand, every snowflake, every individual is unique. As soon as we forget that, or think in terms of "the people," the party, the mass, we have a whole different culture. Mass culture presents us not with life but antilife, the sizzle instead of the steak. It thrives on disinformation, falsies, shadows, moral evasion, and phony fellowship.

The individual is active, the mass is passive. The individual is the arrow, the mass is the target. Individuals such as musicians and writers show and tell. The mass waits to be shown and told. Get into millions and you are in the mass.

Of course individuals must live in community and share things with friends and neighbors. Common culture is the glue that has always held people together—the sense of being someone in some place at some time, the sense of belonging. Until we recover that, poor Humpty will be isolated and frustrated, facing a life full of sound and fury, signifying nothing. Shooting, shouting, and marching will not solve the problems.

Just what is common culture? Where did it come from? Are we losing it? Can we get it back?

Common: shared, joint, united; belonging to all alike; held in *common.* "Longing the common light to share" wrote John Dryden in 1697. We first got the light when we became *Homo sapiens* and banded together to survive in a hostile world. People come together to be together; to love and be loved. To do this they develop language, gestures, codes, symbols. But the most important "thing" they have in common isn't easily defined. Rousseau called it "the general will," Herder the *volksgeist,* Christians the "Holy Spirit."

Many uses of "common" carry forth this concept. A common denominator works for all numbers; common law applies to all; a common right is the property of every individual. As Americans we have inherited centuries of communality and attempted to codify our basic beliefs in a Declaration of Independence and a Constitution. Ours is the oldest written constitution in the world still functioning. Common culture finds its home in community, a social group of individuals who share a common heritage, and fight to defend it. We are, as Aristotle pointed out centuries ago, social animals. We do not come together just to do things but to be human.

Culture and character grow in the same soil. Landscape and memory are Siamese twins. If we don't know who we are, and where we come from, we give up our personhood. In Greek mythology, Antaeus lost his great power and was slain by Hercules, who lifted him off the ground, out of touch with the earth, which was the source of his power. Out of touch, out of mind, out of community.

The rupture, even breakdown, of community in our nation is our overwhelming major concern, for that breakdown sounds the death knell of common culture and finally of humanity itself. Loneliness and alienation are all too often the hallmark of our art, literature, and daily life. Crime, violence, and terrorism grow at an alarming rate; we find ourselves voting for jails instead of schools. Polarization divides blacks from whites, males from females, rich from poor, children from parents, labor from management, liberal from conservative.

But the most alarming gap is between the have-mores and the have-lesses. As production has gone up, workers' wages have declined. The income of the top 1 percent of the population, who control 20 percent of the wealth, rose 78 percent between 1977 and 1989. Ours is a society of diverse economic groups, suspicious of both the future and one another.

What of the poor, the jobless, the homeless? the Pope asked. How long can this inequality continue? Was he implying that Humpty Dumpty might be headed for a fall? Did we not know that in 2002 our social health index hit an all-time low in six categories: children in poverty, child abuse, average weekly earnings, health insurance coverage, health costs, and the gap between rich and poor?

Popular culture mirrors our growing discontent and division. Pop music is in a style free-fall, crunching and bunching odd parts and snarling at a culture that tolerates and supports it. Gangster rap urges us to kill cops and makes light of rape and violence. Some months ago we had a group in our region called Southern Culture on the Skids. The group was well-named, and those of us who are proud of our Southern heritage were appalled. A local music critic summed up the performance: "low-down-dirty-ugly-thunder that's straight out-of-the sewer blooze." The lead singer, it seems, had been inspired by watching bugs.

Is that how to renew our common culture? Hardly. There is no simple way, no quick fix, no magic potion. The renewal may be helped, but it cannot be effected merely in Washington, Richmond, or New York. It must take place in the human heart one heart at a time. We have met the enemy, and he is us. If we want to see where the problem begins, we have only to look in the mirror.

Then, one by one, we must decide to renew ourselves and our communities as we move swiftly into a new century and millennium. No easy task, this. Enormous physical and psychic changes have destroyed ancient truths and attitudes. Leading intellectuals and writers, calling themselves existentialists or deconstructionists, deny the possibility of solutions to our pressing problems; they even deny the value of language. Others downgrade Everyman, once the hero of epics and morality plays, to Subman: a finite clod unmoved by a spark. This will never do. The time for reconstruction and renewal has come. We must lift up our hearts, reviving the Great Tradition which sustained us for centuries. We must master our new technology, lest it master us. We must learn to separate information from knowledge, and knowledge from wisdom. With lots of pluck and some luck we might be able to rescue Humpty Dumpty before he falls. We may not get a second chance. And we must do it one by one.

T. S. Eliot merely repeated what great thinkers have known for centuries: "In my end is my beginning."

"Ends" are much on our minds these days. We have just ended both a century and a millennium, and (according to some) modernism, heroes, nationalism, colonialism, WASP domination, Old-Boy networks, defiant individualism. New items for the list pop up daily.

This has created much new tension, confusion, and controversy, all of which is reflected in the people's culture—popular culture. A vital question engrosses us: *What next?*

What Lies Ahead?

Some future day when what is now is not
When all old faults and follies are forgot.

Arthur Hugh Clough
Poems and Prose Remains (1869)

As the new century begins, a new gold rush is on. The prize isn't the glittering metal from the earth, but the silicon chips in "Computerland."

The old capitalism and new technology? *That* is the question. We are sweeping the globe, and we are told we will soon live not in nations but a Global Village. If the boom doesn't go bust, we may indeed have a global *economy.* But will we have a global *society?*

When we look for signs of what awaits us in cyberspace, we realize that what we have already got, in addition to mountains of undigested facts, is megahype, the vastly optimistic exaggeration about the power and wizardry of computers. Who invented the hype? Mainly people in the computer business, or those linked to academic programs financed by them or their military customers. Upbeat and bullish, these data merchants had one supreme mission: selling. Soon they blanketed our culture and our campuses like the morning dew.

Meanwhile we are lost in a jungle of information, beset by statistical blizzards that numb and befuddle. Computer glut has become a strategy for social control, a way to dazzle and disarm us. Politicians and promoters everywhere (hiding behind the unassailable cover-up of the "The computer says . . .") have profited and prospered via computers. A new class of software millionaires has sprung up. Recall Big Brother in George Orwell's *1984* sending out endless statistics on production and consumption. Recall too how it was by computer statistics (some later disproved) that Ronald Reagan sold the nation on his Star Wars program.

When Theodore Roszak published *The Making of a Counter Culture* in 1986 he noted that there were 2,200 databases on the market.

I do not know how many more there are now, but I do know that they have different codes, protocols, and command languages, and we have our own Tower of Babel.

Some Japanese have begun to send timeless messages, adjusted to our new problems. Here are samples, which suggest how things might change:

> The Web site you seek
> Cannot be located, but
> Countless more exist.

> Chaos reigns within.
> Reflect, repent, and reboot.
> Order shall return.

> Program aborting:
> Close all that you have worked on.
> You ask far too much.

> Windows NT crashed.
> I am the Blue Screen of Death.
> No one hears your screams.

> Yesterday it worked.
> Today it is not working.
> Windows is like that.

> Your file was so big.
> It might be very useful.
> But now it is gone.

> Stay the patient course.
> Of little worth is your ire.
> The network is down.

> A crash reduces
> Your expensive computer
> To a simple stone.

> Three things are certain:
> Death, taxes and lost data.

At 9 a.m. on February 1, 2003, the news flashed around the world: COLUMBIA DISASTER. The space shuttle Columbia, proud veteran of America's triumph in space, disintegrated over Texas, just sixteen minutes from a scheduled landing in Florida. NASA lost contact and data when Columbia began her reentry at 207,135 feet, gliding without power at 13,000 mph.

All seven astronauts were killed, as debris rained over 100 miles, stretched over much of Texas and part of Louisiana. The crew never had a chance: two women and five men, including an African American, an Indian woman, and an Israeli test pilot. The fallen crew members spanned the globe of diversity. That heightened the event's global significance and impact.

Space flight had become routine over the decades; space was no longer just for dreamers but also for bean counters. We had even built a space station that could take up tourists and experiment with things like slime mold. We thought we had "conquered space"; that we could go ever deeper into the dark unknown, until we could prove that we were masters of the universe.

Not so. Columbia TV footage showed a bright light followed by white smoke plumes streaking across the sky. There was no escape, no solution. The fierce outer atmosphere of Earth has powers that surpass ours. There is nothing routine about space shuttle travel.

What now? What next?

Recalling a space disaster in 1986, the enormous cost of a space program plagued by problems, and President Bush saying "The cause in which they died will continue," skeptics raised tough questions. Was this high-flying case of technophilia, of Buck Rogers in the Sky, practical and proper in our world? Some wondered why monkeys or machines weren't adequate payloads. Weren't the space treks actually acts of nationalistic chest-thumping? Should we risk human lives, and endless funding, to bring back a few rocks from the moon?

Perhaps Rick Tomlinson, founder of the Space Frontier Foundation, summed up the whole program best: "You go up, you float around, you see the blackness of space, you throw up, you come down and get a T-shirt."

Many others say we must continue. The army of benefactors from space programs insist that we labor on, since space is the last place where humanity might finally "get it right." Just when, if ever, will that be? And can we "get it right" in our space on earth?

The tantalizing question of "What lies ahead?" intensified in the frantic search for answers. We may never know for sure what happened to the Columbia, or why, but as technology becomes more complex, tiny failures can lead to catastrophies. Columbia had 2.5 million parts, with 27,000 tiles to protect against the intense heat. Success of the flight depended in part on computers, where a single "1" instead of "0" can mean the difference between landing or crashing. One false touch of the finger, and . . .

When we extend our reach, going from technology to technopoly, in which machines control humans, uncertainties and complexities multiply. We end up on the edge of chaos and tragedy.

Have too many lives and dollars already been spent on this venture? If experiments in space must go forward, how shall they be done? Is sending people aloft not only unnecessary (we could give the job to robots) but incredibly expensive, draining money from essential life problems on earth?

Are we building a brave and better new world, or moving into one that will endanger the one we now have? Stay tuned.

And what about Big Brother reading our e-mail, CEOs robbing us blind, mobs rioting against globalization worldwide, an Axis of Evil with nuclear capability? What happened to "Morning in America," optimism and progress? Are we on top of the mountain or hanging on to the cliff with our fingertips? Have we left the Age of Aquarius for the Age of Apocalypse, as many fundamentalists believe? Can we stop the spread of deadly diseases that claim millions of victims, and doomsday biological and chemical attacks? Does the twenty-first century promise to be Apocalypse Soon? Defense Secretary Donald Rumsfeld thinks a nuclear attack on the United States is "only a matter of time," and President Bush warns that "one canister, one crate slipped into this country could bring a day of horror like one we have never known."

Such fears, once on the fringes of society, have entered mainstream American politics and popular culture. More and more, American foreign policy seems based on trying to block rogue attacks and loose nukes. Books about Armageddon top our best-seller lists; Hal Lindsey's book *The Late Great Planet Earth* became a best-selling book of its decade. Is the Age of Apocalypse dawning?

In our frantic scramble to get the news of the hour on the hour, have we overlooked or ignored the long-range questions that will be the hallmark of our age? Will some of those questions involve what was implied in the Columbia Disaster?

Human nature being what it is, we look for a scapegoat. OPEC? The unions? Terrorists? The bureaucrats? The oil companies? How about Congress? The Moral Majority?

There are those who like to "blame it all" on the hippies, groupies, and draft dodgers. Already thirty when the cry was "Don't trust anyone over thirty," they have been smarting ever since. By now, it should be clear that the 1960s were our last fling, that we were living not only on borrowed time, but borrowed resources, and that both were running out.

Not that the things themselves, or the uses to which we normally put them, are evil. The danger is we surround ourselves with things to ward off boredom, loneliness, and even community. "Things are in the saddle," Emerson observed, "and ride mankind."

What might the "Energy Crisis" mean to America? It might remake our whole image of ourselves, from the cannonball to the rubber ball. The cannonball knocks down everything before it, having the power and energy to do so. The rubber ball moves until it hits something too hard to penetrate; then it bounces back. Bouncing has resilience, humor, perspective. Winnie-the-Pooh realized that when he spoke admiringly of Tigger:

> But whatever his weight in pounds, shillings, and ounces,
> He always seems bigger because of his bounces.

America not only desires, but demands, changes. That was a clear message of the 1980 elections. Many New Deal ideas, which served us so long and so well, must be revised or rejected. We must listen carefully and continuously to what is being said both at home and abroad. Some retreat and reform is inevitable. Yet the major movement of this generation need not be retreat. It can be rebounce. Who knows how high we might bounce next time? We will seem bigger because of our bounces.

Technology controls and dominates our lives. In many parts of the world, however, high tech has made few inroads. At top, a brick worker in India breaks bricks with a primitive hammer for a living, and, at bottom, Indian women weaving material for the American market are doing what people in India have done for centuries. How can these differences be reconciled?

Petite Probes

Surgeons are master probers. They have many knives; some for deep probes, which might involve a heart or kidney; others to dig out a splinter or a wart. One definition of the noun *probe:* "a slender surgical instrument for examining a cavity." Another, of the verb *probe:* "to make an exploratory investigation."

Writers probe too, hence my title. Having attempted larger and deeper probes, I shall conclude with petite exploratory probes. What is petite? The French define "petite" as small of size; tiny. They are masters of petits fours—small tea cakes, variously frosted and decorated. Think of that term now; my "petite probes" are meant to frost and decorate my book. I hope you find them tempting and tasty.

ELVIS LIVES!

Death elevated Elvis from fading legend to persistent myth.

Richard Harrington

Elvis was widely celebrated in 2003, twenty-six years after he died. Or did he?

Say his once-strange name anywhere in the world, and eyes will light up. Elvis! A name that tests your gag reflex. In our new century, first-name-only stars are everywhere, but no name evokes like that of Elvis. He has moved from death to mythology.

Like the gods and heroes of old, Elvis offers multiple images, for he is a psychic chameleon. The first person to sense Elvis's power, Marion Keisker, said Elvis was a mirror. Whatever you were looking for you'd find in him.

Elvis's June 1953 conversation with Keisker, office manager for Memphis Record Service, catches the myth at its very beginning. When Keisker asked the teenager what kind of songs he sang, Elvis replied, "I sing all kinds." Whom did he sound like? "I don't sound like nobody." Before long, everyone would try to sound like Elvis. Keisker pressed Elvis's first pop ballad ("My Happiness") for $3.98 plus tax, and the game was on. The ensuing story has been told time and again, by me and countless others. It's the old Horatio Alger scenario: from poverty to riches, from Mr. Nobody to the King of Opportunity, the Prince of Possibility.

In 1995 scholars from all over the world came to seminars at the University of Mississippi to "examine the Elvis phenomenon." A panel of learned experts tried to answer such questions as "How did he bridge the Black Blues and the White Gospel?" Elvis never talked about "how he did it," and he never showed the slightest interest in the university. But universities have been making big bucks from him, like many other places where he lived, loved, or worked.

The big tourist attraction brings scores of buses, lining up in front of Elvis's Memphis home, Graceland, and visitors end up like Elvis—"all shook up." When Elvis died, the Graceland estate was valued at one million dollars; the revenue from visitors in 2003 topped sixteen million dollars, according to Graceland officials.

That figure was dwarfed by returns from Elvis's recordings. He remains the world's best-selling artist. People all over the world still line up to buy Elvis records, tapes, and discs. No one has ever equaled him: a billion records and singles and a "Young Elvis" stamp by the U.S. Postal Service which sold 500 million copies. Now, that's first class! Thousands of people "impersonate" him. It has, in fact, become a growth industry. Elvis is everywhere and his appeal is endless.

All this for a poor boy born in a shack outside Tupelo, Mississippi. His twin brother died at birth. Elvis was what native Southerners call "poor white trash." Look what he became! Poor no more!

What do we remember? The voice, the songs, the look, of course. Also the breathtaking gyrations, the wild abandon; but then he stopped, smiled, and was a fresh-faced shy young guy, just in from the country, trying to find a job and to make us feel good.

He got into the army, refused to be exploited as an entertainer, but served instead as a grunt with the other guys. They loved that, and loved him. He came out a sergeant.

After that came a long series of hit records, and some "love me—hold me" movies that were hardly memorable. But all the directors and promoters found him easy, even fun, to work with. He always knew his lines, and millions of teenyboppers swooned.

Later on Elvis couldn't resist the fast track at Las Vegas and somehow got into drugs. He gained weight and forgot his lyrics. But he was still Elvis, and broke all records. The audiences came and applauded, come rain or shine. Then, the long good night.

Were he alive in 2003, he would be sixty-eight, not able, we guess, to belt out "Blue Suede Shoes," or swivel with "Jailhouse Rock." Never mind. Don't tell us, as CNN did, that he died with ten different pharmaceutical drugs in his bloodstream. Show us the incredible smile of the most famous stand-alone singer in history. Let us recall that innocent young kid, that Lancelot come forth to defend us, singing "Love me tender, love me true." We do, and, yes, we will.

Dozens of books have been written about every aspect of his life, legend, triumph, tragedy, and death (the bloated, naked Elvis was found dead on his bathroom floor in Graceland, in a pool of vomit). Newspapers all over the world agreed with the much-quoted remark by Bruce Springsteen, that Elvis whispered a dream in everybody's ear and we all dreamed about it. We still do. Now we can get an "authorized" new book called *Elvis: A Celebration: Images of Elvis Presley from the Elvis Presley Archive at Graceland* (2002) (DI Publishing, $50). Edited by Mike Evans, it was produced by Elvis Presley Enterprises (EPE), and brings more and more dollars into the prosperous Elvis Presley Trust and estate. Some other recent Elvis titles speak to the scope of the new mythology. Kim Aldelman has given us *The Girls' Guide to Elvis: The Clothes, the Hair, the Women, and More!* (Broadway, $12.95). The "more" includes Elvis's favorite recipes, including one for Fried Peanut Butter and Banana Sandwiches.

We will never run out of new aspects of the Elvis story. In 2002, Jonathan Goldstein and Max Wallace dug and dug until they uncovered Elvis's Jewish roots. It's all explained in *Schmelvis: In Search of Elvis Presley's Jewish Roots* (ECW, $17.95).

I was lecturing on the myth Elvis to a university class recently, when a hand went up in the back of the room. "You don't know the half of it," the earnest young man said. "We all know Elvis is alive. He turned up at a bar downtown last Saturday night, and sang two songs." "And after that," a young lady with big blue eyes added, "he went on to some more numbers at the Christiansburg Friendly Bar. It was cool!"

Who can top that?

THE TALE OF TWO ENZAS

This is the tale of two "enzas," a great ship, and the way history and popular culture catch up with all of us.

Begin with *influenza*—an epidemic formerly attributed by astrologers to the *influence* of heavenly bodies. We now attribute it to a deadly virus; we take our flu shots and hope for the best. Fortunately, we have vaccines for several types of infection, but not all. Look what happened to the world in 1918.

A television documentary in PBS's *American Experience* series retold the shocking story. The influenza epidemic struck without warning, killing indiscriminately. Scientists brought forth no cure. Millions died around the world, over 600,000 in the United States alone. That's more casualties than we suffered in all our twentieth-century wars. It was especially devastating among troops training and serving in World War I. The disease passed like lightning among close-quartered soldiers and sailors. Some who reported sick in the morning were dead by nightfall. A national shortage of coffins occurred. The epidemic finally passed, having devastated the nation.

If it strikes with that ferocity again, do we have an effective vaccine? Does this help explain why the dreadful epidemic of 1918 is usually ignored in history books, even forgotten in the nation? Will other viruses, such as those for AIDS and Ebola, prove far more deadly than that for influenza?

Over eighty years later we have another fast-spreading epidemic, involving millions, with no apparent cure. The symptoms are quite different from influenza; the victims live believing that we can buy our way into salvation. They are like the little pig who built his house of straw: Who's afraid of the big bad wolf? Sociologists call this new "enza" *affluenza.* The symptoms are reckless spending, material glut, credit card capers, online sprees, mall madness. All races and both genders, many under thirty, are infected. Stock options, big bonuses, and tricky buy-outs fuel a new epidemic of *affluence.* The stock market rose higher and higher in the 1990s. Would it go on forever?

We heeded the 1996 Pepsi slogan—"Buy Pepsi, Get Stuff." We got stuffed. If at first you don't succeed, buy buy again. We were driving down life's highway with the throttle wide open. Get aboard or get off the road!

Historians add a cautionary note. We've been down that road before. *Affluenza* struck America hard after the Civil War, creating what V. L. Parrington called "The Great Barbecue." Robber barons, grabbing up land, railroads, steel, and oil, had a gargantuan feast. More food was spoiled than was eaten. All the "important people" were invited. There wasn't enough room for everybody; there never is.

But something ominous lurked on the horizon: the depression of 1893. Even though it was devastating, we didn't get the message. An even greater setback awaited the nation.

The boom-to-bust cycle repeated itself. After World War I the market zoomed, gangsters gunned, investors bought on slim margins, bankers made huge profits, and the good times rolled. Scandals in Washington were swept under the rug. This was the Jazz Age, when speakeasies flourished and flappers flapped to the Charleston—right up to the Great Depression, when the bubble burst. Starting in 1929, the Depression lasted for a decade. Songs of the period were "Brother, Can You Spare a Dime?" and "Can I Sleep in Your Barn Tonight, Mister?"

How does all this help explain the phenomenal popularity of the blockbuster Hollywood movie, *Titanic*? The story of that worst-of-all maritime disasters has been told and retold. The largest and most luxurious ship afloat struck an iceberg off the Grand Banks of Newfoundland. Thought to be unsinkable, it sank quickly, carrying 1,513 people to their deaths. Why is the *Titanic*'s story so gripping in the new millennium? Why tell it again?

Because it is both a parable and cautionary tale. The *Titanic* symbolized the affluence and unbounded optimism of the early twentieth century. The Greeks would have called it our *hubris,* belief in our infallibility. When the ship was launched in 1912 we were enjoying the closing years of the Belle Epoch—those forty years between 1870 and 1914 when we experienced no major war, widespread prosperity, and an explosion of new technology, literature, and art.

We passed quickly from the Age of Steam to the Age of Electricity, with Edison's incandescent lightbulb brightening our future in 1879. Then, in quick order, the phonograph (1880), synthetic fiber (1883), electric motor and Kodak box camera (1888), diesel engine (1892), Ford car (1893), X-ray, movie camera, and radio telegraphy (1895), powered flight (1903), Special Theory of Relativity (1905), the assembly line (1908), new styles of poetry and painting (1908). In 1913 the French writer Charles Peguy noted that the world had changed less since the time of Jesus Christ than it had in the past thirty years.

But the *Titanic* sank, the Belle Epoch ended, and World War I proved to be the most devastating war in centuries. What happened? History happened.

Of course we neither want nor expect another *Titanic* disaster or World War, although violent outbreaks in Asia, Africa, and the Middle East raise the possibility. What the historians might say now is be cautious; learn from past mistakes. Why did the great Spanish Empire collapse in the seventeenth century, the French in the eighteenth, the British in the twentieth? They all became too greedy, self-confident, and aggressive. They thought they could patrol and control the world. They were wrong.

Suddenly we face military and financial dilemmas all over the world which hearken back to earlier overextensions by ambitious powers. Some of history's greatest tacticians—Genghis Khan, Alexander the Great, Julius Caesar, Napoleon, Stalin, Hitler—bit off more than they could chew. History spat them out. Today the United States is the sole remaining superpower. No one knows how long this will be so, or what we will do with this power. We must lead. We must be bold. But *how* bold?

Is this the time not for the green light of boldness but the yellow light of caution? Perhaps more caution would have prepared us for the precipitate economic turnaround at the end of 2000, when a million new millionaires ended up with a million pieces of near-worthless paper. The obvious moral of our story?

History happens. It will happen to us.

SUCCESS IN AMERICA: PLUCK AND LUCK

> I will succeed, for my talents and talons are strong.
> By pluck and luck, I will reach the top.
>
> Horatio Alger

In America success is a god, and Horatio Alger remains one of its most popular and pervasive priests. His own life story contains the elements of success, leading to bittersweet failure, which is with us today, as the mighty CEOs of billion-dollar corporations such as Enron and WorldCom are marched off to jail in handcuffs.

Just who was Horatio Alger? How did his hundred-odd novels shape the imagination, and live on through numerous modifications and transformations? To find answers to those questions, turn to a vanished world, the foundation of the one we live in now.

Empires share artifacts, mentifacts, icons, heroes, myths, gestures, rituals, and symbols. Christians spoke of the Holy Ghost and the native American people who flourished centuries before Europeans arrived, space of the Great Spirit. European intellectuals still write of the *volkgeist.* Now we deal with corporate capitalism and the "American Dream."

All cultures must cope with change, process, and progress. These three keys to every age help explain America's current stance in the New Age. Our longstanding myths, Success and the Self-Made Man or Woman, still grip us. The motto is "Onward and Upward." Our proof came from Horatio Alger Jr. (1832-1899). Of old Yankee stock, he believed in chastity and stern morality and yearned for success. While at Harvard College he indignantly changed his lodging when his landlady appeared in her negligee. "I might have seen her bare, but I did not look," he wrote. Later he journeyed to Paris, where he did things that would have shocked his ancestors. He looked. These two lines in his diary document the fall from grace: "I was a fool to have waited so long. It is not nearly so vile as I had thought."

Then Horatio ground out 135 novels that sold over 20,000,000 copies. While others wrote single novels, Alger turned out whole series: *Ragged Dick, Tattered Tom, Brave and Bold, Luck and Pluck, Way to Success, Atlantic,* and *Pacific.* He sometimes wrote two books at once; his potboilers were best-sellers. Even as he wrote, men such as Henry Ford, Thomas Edison, and John D. Rockefeller were proving him right. Alger even wrote a little ditty that was, in effect, his formula:

> Strive and succeed, and world's temptations flee
> Be Brave and Bold, and Strong and Steady be,
> Go slow and steady, and prosper then you must
> With fame and fortune, while you Try and Trust.

But fame and fortune deserted him. Finally he ran out of luck, even out of ditties. "I should have let go," he admitted. "How many times I wanted to!" A violent affair with a married woman, antagonism from his family, and increasing despair found him tipping the Astor House desk clerk to point out celebrities, or perhaps pounding the drum in newsboy parades.

Alger finally sought refuge in the Newsboy's Lodging House on a dingy street in New York City. The gilt had long since worn off his style. At the century's end, alone and forgotten, he died quietly in a drab dormitory room.

Years earlier, half in jest, he had written his own epitaph. As Gary Scharnhorst notes in *The Lost Life of Horatio Alger, Jr.,* it tells us a lot about popular culture:

Six feet underground reposes Horatio Alger,
Helping Himself to a part of the earth,
not Digging for Gold or in Search of Treasure,
but Struggling Upward and Bound to rise at last in a New World,
where it shall be said he is Risen from the Ranks.

CURSES, FOILED AGAIN!

No wonder baseball is called "The National Pastime." It's rich in history, fact, fiction, and folklore. There are even baseball curses, and thereby hangs our tale.

Baseball burst upon the national scene in the 1830s. One myth holds that it was invented by Abner Doubleday during the spring of 1839 in Cooperstown, New York. Visitors still flock there today to visit the Baseball Hall of Fame and Doubleday Field. But many more crowd into stadiums nationwide, where millions of people spend millions (perhaps billions) of dollars to see The Old Ball Game and to yell, "Kill the umpire!"

Some historians say the game was an offshoot of the English game of rounders; in America, the game has been known as roundball, townball, and of course, baseball. The first league was formed in 1857 and the game prospered during the Civil War. The Cincinnati Red Stockings became the first professional team in 1869. By then there were more than 400 baseball clubs in the nation.

Over the years new teams, new fans, and new heroes swelled the baseball scene. The game is more than its stars and winners, more than Babe Ruth's prodigious homers, Ty Cobb's stolen bases, Bob Feller's fastball, or Willie Mays' classic basket catch. It is engrained in the American psyche; some would say it's a kind of religion.

Along the way there have been scandals, heartaches, and yes, curses.

Baseball's curses became the big stories during the playoffs in 2003, when they seemed to be working in full force. Enter the Boston Red Sox and the Chicago Cubs. The Red Sox hadn't won a World Series since 1918, and the hapless Cubs hadn't won a Series since 1908. Was that about to change? Polls showed that fans favored a World Series between the Red Sox and the Cubs. Curses, foiled again! The New York Yankees and the Florida Marlins played the Series, and Boston and Chicago went into mourning.

Boston's curse and woes began when Babe Ruth (The Bambino) was sold to the New York Yankees. Why did Red Sox owner Harry Frazee sell his ace player for $125,000 in 1920? Just eighteen months earlier Ruth had led the Red Sox to a World Series title, ironically over the Chicago Cubs. And the Red Sox had won five World Series titles previously (1903, 1912, 1915, 1916, and 1918). Some say Frazee needed money to finance his girlfriend's Broadway musical, others say that Ruth was unhappy with his $10,000 salary with the Red Sox, and that he no longer wanted to pitch. Whatever the reason, when the sale occurred, The Curse of The Bambino was born. The Yankees have won twenty-six World Series titles since that historic sale, the Red Sox, zero.

The Red Sox have won the AL championship and have played in four World Series games since 1918, but in each of those events, the Sox lost the Series in Game 7. Boston fans cannot forget the fateful Game 6 of the 1986 World Series against the New York Mets, when Mookie Wilson's grounder rolled through the legs of first baseman Bill Buckner, giving the Mets the win and forcing Game 7, which the Mets won.

Curse reversal has been tried. Two female Red Sox fans attempted to end the curse using *feng shui*—the Chinese art of bringing forces into harmony. Outside Fenway Park during the 2003 playoffs, they rattled chimes and burned incense and a picture of Babe Ruth. It didn't work. Boston blew a 5-2 lead in the eighth inning of Game 7, and lost in the eleventh inning when Yankee Aaron Boone hit a home run off knuckleball pitcher Tim Wakefield, who was considered "impossible to hit." Curses, foiled again!

The Chicago Cubs curse is even more fascinating. Fans call it "The Billy Goat Curse." They can date it precisely—the fourth game of the World Series on October 6, 1945, at Chicago's Wrigley Field. The Cubs had already won two games out of three in Detroit, and the future looked bright as they returned to their home diamond. But wait.

Up to the admission gate strode a Greek tavern owner known as William "Billy Goat" Sianis, who attempted to bring his billy goat, Murphy, in to see the game. He bought the goat a ticket, but the animal was denied admission. The fans and the ushers complained that the goat was "too smelly." Did P. K. Wrigley, the Cubs' owner, refuse the goat's admission? Apparently the goat had been admitted to previous games. The angry Mr. Sianis cursed the Cubs in broken English, saying, "Never again will World Series be played in Wrigley Field." It never has been. The Cub lost the 1945 World Series to the Detroit Tigers (again, in Game 7) and have not been in a World Series since.

The curse seems to linger on despite attempts to lift it. Sam Sianis, nephew of "Billy Goat" Sianis, and current owner of the legendary Billy Goat Tavern in Chicago, tried to bring a goat to the park again in 1973, and again the goat was refused. Sam and his son have come to the park with goats four times since, in 1984, 1989, 1998, and in 2003. All four years the Cubs lost, twice with the team just one win from entering the World Series.

Witness the 2003 NL playoffs against the Florida Marlins. Going into Game 6, the Cubs were leading the series 3-2, and had home field advantage. All they needed was one win and they would be World Series bound. The celebrations were being planned. The Cubs were ahead 3-0 going into the eighth inning and with Mark Prior pitching they were five outs away from their first NL pennant since 1945. Then the bottom fell out. A foul ball that supposedly would have been caught by Cubs fielder Moises Alou was instead deflected by a fan, and their shortstop missed an easy grounder. The Marlins racked up eight runs in the eighth inning of Game 6, tying the series at 3-3 and forcing Game 7. Had the curse kicked in once more?

The Marlins went on to win Game 7, 9-6. Another World Series was denied the Chicago Cubs. Said Marlins manager Jack McKeon,

"The Cubs were America's favorite. I think we're the darlings now." Cubs manager Dusty Baker philosophized, "We didn't lose the pennant; the Marlins won it." Baker's four-year-old son injected youthful optimism: "The Cubs will win next year," he said.

Will they? Or will the curse foil their chances yet again? All the thousands of Boston and Chicago fans can do is wait and hope. Why should they believe in curses?

And yet . . .

FEAR

From long-legged beasties, and things that go bump in the night
Good Lord, deliver us.

Book of Common Prayer

An old virus has taken on new power in the twenty-first century: *fear*. Things are going bump not only in the night, but all day long.

Consider a simple, once pleasant task: opening the daily mail. Most of it was once first class, with a bit of junk thrown in. Now it is mainly junk, with an occasional first-class letter or bill included. And what is the motive of much of the junk? FEAR.

I give you examples, from today's mail. I quote:

- Dear Friend: I don't want to alarm you, but there is cause for concern. Perhaps there is a "Silent Killer" lurking in your furnace. Inspect before you get sick—OR WORSE!
- Do you know if that over-the-counter drug is safe? Some of them might kill you!

What has brought on this rash of irrational fear? A simple answer pops up: 9/11. Polls show that two-thirds of New Yorkers expect another surprise attack. Carbon monoxide can escape from your furnace, and certainly some drugs do more harm than good. Thus has it always been. But has a newer deeper factor emerged? I think so. Let's call it techno-fear.

Is this deep unsettling fear irrational? Perhaps so—but not to those who feel it. Statisticians can claim that your chances of being attacked by a sniper in the Washington area (where the famous attacks occurred in 2002) are infinitesimally small—about 1 in 517,422—but does it help to know the sniper risk is small compared to the risk if you drive, smoke, or jaywalk? Not at all.

Fear is an ancient and genetic imperative, flaring up at certain times, and is quite capable of overcoming reason. We live in one of those times.

How did terror change our culture and lead us into a new, uncharted, and dangerous war? No one can say. Fear stalks the land, and a ruthless sniper could aim his gun at an innocent unsuspecting victim. Cells all over the world are plotting on how to bring violence and destruction to America. Hate hides in the back alleys, and ignorant armies prepare to muster and march by night. We watch the news and spend nights wondering if we locked all our doors. Listen. What was that noise?

Fear is one of the complex factors that is always present, and which helps to shape our popular culture. Just the stress of fear is dangerous. We must be afraid of being very afraid. President Franklin D. Roosevelt was right in 1932 when he told us we have nothing to fear but fear itself.

SATURDAY MORNING SHOOT-OUT

"GOTCHA!"
MASSACRE TRIUMPH

Saturday morning at last, Mom and Dad can turn off the alarm clock, ignore the e-mails and stress pains, and pretend they have a mini-vacation. Not so with the kids: this is the moment they have been waiting for all week. It's shoot-out time. They rush to the TV and join the fray: shoot-'em-up science fiction, blood and guts, bold beautiful heroes, scary supernatural villains. Maybe even Sponge-Bob versus Ultraman! And who makes all this possible? That genial Master of Magic, Alfred R. (aka King) Kahn.

Perhaps you don't know about the king, Alfred Kahn. He has played a big role in kid pop culture, having introduced Cabbage Patch Kids and Pokemon to an eager nation earlier. He is worth probing. In January 2002, he leased four years of the preteen Saturday morning time slot, 8 a.m. to noon, for $101 million. He's been getting richer ever since. Long live the king!

Chief executive officer of 4Kids Entertainment, Inc., Kahn released his new line in September 2002, and everyone could see what his target was: boys. He explained that when they fantasize, they want power, strength, and speed: Pow! Kazam! Voom! And that's just what he intends to give them. Welcome to the newest new look in popular culture.

His new line-up, called Fox Box, shows what he has in mind. Splat! Voom! Gotcha! He is out for the preteens, giving them more blood and gore with Japanese imports; and he's into a twenty-four-hour-a-day production schedule to do it. He makes no apologies for his tactics.

Encouraged by this, Kahn expanded his Manhattan studio, connecting the eleventh floor of three adjacent buildings to accommodate some ninety sketch artists. When the shows arrive from Japan, the staff must produce translations, dub the films, and compose music made for American, not Japanese, ears. All this takes much talent and effort. Norman Grossman, who heads the production unit, says he is trying to give the imports a lighter touch. He sometimes uses campy humor. In one scene the hero says the monster he fights is uglier than his mother-in-law. "Yes," the hero's sidekick responds, "but her nose is bigger." Then on to the climax in which the monster villain explodes into bits.

What are some of the new superheroes he will feature? One is Ultimate Muscle, in which the good guy saves the galaxy; another is Ultraman Tiga, in which the Mighty Morphin Power Rangers clean things up. Fighting Foodons and Kirby bash hideous monsters, while Kahn's round-the-clock staff do new versions of the once-popular Teenage Mutant Ninja Turtles.

There is more now to woo boys into buying action figures, video games, game cards and new tie-ins, all of which bring big royalties.

Kahn thinks big and he is big. Over six feet tall and weighing 208 pounds, he lifts weights every morning at 5:30. Bench pressing 500 pounds, he is both the oldest and strongest of his 160 employees. And he's ready to roll. He wears an electronic police beeper that alerts him to high-speed chases and major crimes in New York. He is a kind of action hero in his own right. He always wanted to be a cop.

THE ZIEGFELD GIRLS

There is nothing like a dame.

South Pacific

"Thank heaven for little girls," Maurice Chevalier sang in *Gigi,* while Pat Boone was belting out "Venus in Blue Jeans." Songs, poems, shows, and films about the American girl (females remained "girls" for decades, just as blacks remained "boys") have been the bread-and-butter of popular culture and music. After all, "A Pretty Girl Is Like a Melody." Feminist protest has dampened the "girl" label, but not the mystique. Could it be that "girls" have become, in our secular world, icons?

The very idea might offend those who associate icons with the church, and the dominant figures of the past. But just what is an icon? An expression of internal convictions, cultural ciphers that operate on an emotional level. By this definition, have not the computer, satellite, TV set, even the Coke bottle become "iconic" in 2003? Doesn't the Coke bottle mirror the curves of the female body?

Move to Florenz (usually called Flo) Ziegfeld and his "beautiful girls." Where did the term "show girl" come from? Who decides how the "All American Girl" should look, dress, and act? We might find clues in many generations and places, but I believe one man is a prime candidate for putting today's "girls" on stage and in the limelight. He bridged two centuries (1869 to 1932) and created what Randolph Carter described in *The World of Flo Ziegfeld* (1974). He became what is known as a Broadway Legend. That "world" has special significance for popular culture.

In her book *Ziegfeld Girl: Image and Icon in Culture and Cinema* (1999), Linda Mizejewski notes that for Ziegfeld, choosing a beautiful young girl (usually blonde, slim, with small lips and a straight nose) was only the beginning; then the hype and humbug were applied. Women couldn't possibly be as beautiful, desirable, and breathtaking as that. Yet they were, under the Ziegfeld master touch of illusion.

Marketing, rather than invention, was his genius. He had a formula for show business: one-third glamour, one-third merit, and one-third advertising. Properly blended, they were almost sure to work then as they do now.

Ziegfeld was a choreographer of racial, sexual, class, and consumer desires that were deeply rooted in American culture. He turned away from the loftier ideals of his father, who presided over the Chicago Musical College from 1876 to 1916, and turned to the earthier popular entertainment. His "Ziegfeld Follies" were staged nearly every year from 1907 to 1931. They featured two things central to the American hype: huge budgets and outrageous one-upmanship. They became, in the phrases of the day, a "national institution." Faces and titles change, but the institution, like "Old Man River," just keeps rolling along.

"NEW EUROPE" GOES POP

In 2003 America got a new Gulf War II and Europe got a new split. Those opposing the war were labeled "Old Europe," mainly Western Europe, including France, Germany, Belgium, and some Mediterranean countries. "New Europe" encompassed Eastern Europe, including countries freed from Soviet control—the Balkans, Hungary, Bulgaria, and Poland.

Not only was diplomacy and politics involved, but also popular culture took immediate note. Cartoonists had a field day with the whole matter, and pop culture too. What fascinated Americans and lured them into war was the chance to reinvent democracy in nations that had lost it, to claim them for "our side" in new struggles. Signs of decay and crisis appeared in "old" nations and the chance of rediscovery and liberation in the "new" America.

New voices and ideas have popped up. One example is Slavoj Zizek, a Slovene trained in France, who made whirlwind tours in America where audiences flocked to hear him. He specialized in American popular culture and used famous movies, such as *Vertigo* and *The Sound of Music* for analysis. He was a breath of fresh air.

"New" Europeans are obsessed with national identification: What elements hold a given community together? Zizek begins much of his work with this question. The bond linking people together, he believes, is a shared relationship toward a thing. This relationship is structured by means of fantasies and the belief that "our thing" makes possible our "way of life." Our "thing" is threatened by "others," outsiders, who can never understand just what our "thing" is. They can't grasp it, but they can threaten it. The national "thing" exists only as long as members of the community firmly believe in "the real thing."

A good example of the Holy Spirit occurs in Christianity. To believe that Christ lives after his death means believing in belief itself. This has been a sustaining factor in the Christian world for centuries, and with others, in other religions too. We see this clearly in our own day. The role of Islamic faith in our new century is a determining factor of Gulf War II.

When I read Zižek, I think of another example: the motto and power of Coca-Cola. After all, "It's the real thing." Coke is available in over 100 countries around the world. With Zižek, we have a new look at Coke from the New Europe.

The Coke bottle is the most widely recognized commercial product in the world; one of the few truly participatory objects around the globe. Usage cuts across nationalities, boundaries, classes, and age groups. Surrounding the Coke bottle are endless anecdotes, personal associations, and behavior traits (such as the Coke Break).

Zižek goes on to expand the physical qualities of popular objects (such as the Coke bottle) into the intangible mythic realms. He believes that a nation exists only as long as its specific enjoyment continues to produce a set of social practices, transmitted through national myths that structure these practices. What is at stake in racism and ethnic tensions is always the possession of the national "thing." The ultimate root of our fear of "aliens" and "others" is not their physical menace but their ultimate motives and intentions, which remain completely impenetrable and unknown to us.

By linking history, ideology, and psychology to popular culture, "New Europe" thinkers are opening up new frontiers and probes.

ARE WE EMPIRE BOUND?

A new buzz is on. When old questions are answered, new ones pop up. Should we go to war in Iraq? We did. Could we win in a month, with incredibly low casualties? We did. Did that raise a dozen baffling new questions and dilemmas? It did.

None is more prominent and pressing than this: as the world's only superpower, with military might that no nation, or combination of powers, might oppose, are we on a roll? Will the now-defunct British Empire find its successor in the once-British colony in which the

British Empire first planted its flag (in Jamestown, Virginia, in 1607)?

Politicians, pundits, journalists, and historians fill the airwaves, print media, TV and computer screens, chat rooms, and barber shops with sure-fire answers. Those who know a lot about empires, and many more who don't, are quick to answer, pontificate, and predict. What to make of all this sound and fury as sides line up for endless verbal combat?

They tend to repeat just what they said on our previous questions. Those who opposed the war are, in the main, opposed to an empire; those who favored it, and savor the sweet taste of victory, favor our action. Though labels are too easy, we still have liberals versus conservatives, hawks versus doves, just as the English had earlier with Little Englanders and Big Englanders. We must look beneath the labels, but first we need some markers. Just what *is* an "empire," and what can historians tell us about the great ones of the past?

"Empire" is derived from the Latin *imperare,* to command, and *imperium,* absolute authority, empire. The concept has centered on an extensive territory or enterprise under single domination or control. The best example might be the Roman empire, which evolved into the Holy Roman Empire and lasted a thousand years. It finally collapsed, Voltaire noted, because it was neither holy nor Roman nor an empire. But remnants remain, especially in the Vatican. Other empires filled the vacuum. Some, such as the first French Empire (1804-1815), are remembered largely for style in furniture and clothing. In women's dress it implied a high waistline with undraped skirts hanging loosely, and the style keeps cropping up in extensive period Technicolor movies made in Hollywood.

The French Empire is no more, and the idea of an American successor appalls Europe, but these words are still embossed on the magnificent home of France's empire-minded Louis XIV at Versailles: "À Tout Les Gloire de la France."

The British empire held on much longer than the French, spanning the eighteenth, nineteenth, and first half of the twentieth century. Under Queen Victoria, who reigned for over fifty years, almost a quarter of the globe answered directly or indirectly to London. May 24 is celebrated as Empire Day, the anniversary of Queen Victoria's birth, observed throughout the British Empire. The popular slogan proclaimed "The sun never sets on the British Empire." It set after World War II.

If we are the heirs of England's empire (which once included what is now Iraq) we might look for similarities and strengths. There are many. First is the English language. An estimated four million people spoke it in Shakespeare's day; now it is the nearest thing to a world-wide tool and passport. Patriotic fervor and chauvinism is pervasive. Count the number of American flags you see since 9/11.

Eighty-six years ago a British army entered Baghdad, and would soon influence other Middle Eastern countries, including Syria, Iran, and the Persian Gulf states. British World War I pro-imperialists such as Mark Sykes and Leo Amery are echoed today by American neo-conservatives and politicians. Some liberal historians, dedicated doves, embedded in academia, are already firing back.

In a new book by Britain's Niall Ferguson, *Empire: The Rise and Demise of the British World Order and the Lessons for Global Power* (2003) he points out many imperial dangers but also accomplishments. He emphasizes the British exploitation of slaves in the eighteenth century; the massacre of aborigines in Australia; the grabbing of land in Africa; and gunboat diplomacy. But turn the coin over. The Empire also pioneered free trade, free capital movement, the abolition of slavery, and free labor. A global peace ensued unmatched before or since, and improved norms of law, order, and governance around the world. The victory in World War II vindicated the empire's existence, Ferguson believes. It was sacrificed to stop Germany, Japan, and Italy from keeping their empires. The story is too complex for glib answers, which we hear from all sides.

In the twenty-first century the United States has the world's most successful economy, as did Britain in the eighteenth and nineteenth centuries. As with the earlier imperials, we can do a great deal to better the lives and futures of inhabitants of less technologically advanced societies. We might even further Thomas Jefferson's famous words in our own Declaration of Independence: we are all created equal, with certain inalienable rights, including life and liberty. Might these bold words now become meaningful to millions who have never dared dream that they might apply to them?

Like it or not, the idea of empire is always present, be it with the Babylonians, Assyrians, Egyptians, Greeks, Romans, Chinese, Ottoman Turks, Japanese, or various Western European nations. History tells of ever-recurring empires.

Ferguson, a Fellow of England's Jesus College, Oxford, believes that globalization is occurring and will expand; America, he concludes, has taken over our role, without yet facing the fact that an empire comes with it.

Some Pentagon planners are "facing the fact," formulating a new operating theory. It centers on "the gap" between the well-operating parts of the world, where globalization is moving forward, the functioning core. Whole areas of the world aren't in it: the Caribbean Rim, much of Africa, the Balkans, the Caucasus, parts of the Middle East and Asia. Most of the two billion people in this scenario are very young, with low incomes, consumed with conflicts and anger. This has created the "nonintegrating gap."

Unless we deal with it, the argument goes, no real world peace or security can occur. We can't eliminate the gap, but perhaps we can shrink it. We must stop the ability of terrorists to access the functioning core, aided by "seam states" along the gap's bloody boundaries. This explains why we have responded to global crises since 1990 and in Iraq, and why, according to the hawks, we must continue to do so. Prelude to Empire?

Whatever we plan, The American Moment is here.

INSIDE THE CAGE

Stone walls do not a prison make,
Nor iron bars a cage.

Richard Lovelace
"Lucasta," 1649

Three centuries after Lovelace wrote these famous lines, we still wonder how to define "prison." Wealthy Americans spend fortunes to live behind gates that keep the world out: self-imposed prisons. In 2003 more than 7 million households have entries controlled by gates, guards, entry codes, and key cards. What has fostered this fortress mentality?

Professor Setha Low has probed this development and raised this question, in a recent book titled *Behind the Gates: Life, Security, and the Pursuit of Happiness in Fortress America* (2003). She tells of a visit to her sister Anna in San Antonio, who lives the gated life there, dressed in cowboy boots, blue jeans, and DKNY T-shirts. She also tells of moving around the house, feeling trapped, claustrophobic, and uneasy. She accidently sets off the burglar alarm. Is all this part of the American Dream or the American Nightmare?

This is what sociologists call "fear flight"—trying desperately to find safety, community, and "our kind of people." Two special kinds of fear dominate: fear of crime and fear of others. Fear of crime is based on the possibility that criminals can "get you." Lurid media reports have created a culture of fear. Hence the wealthy have long lived in separate zones (the "right side of the tracks," in my native Virginia city, Roanoke) and have made restricted land covenants. Today's new gates are merely an expansion of an old residential pattern.

Fear of others is also a very old thing. Franz Kafka describes it in *The Great Wall of China.* Frequent bathers have always wanted to separate themselves from the Great Unwashed. Our new stress on equality and diversity has brought many "others" to our doorsteps. The walls are making visible the systems of exclusion that are already there; new walls are constructed of concrete.

Of course 9/11 and the ever-growing fear of terrorism have greatly added to the urge to put up the gates, close the borders, inspect the immigrants. Homeland Security has become a major part of the new American scene. The nation moves from alert to high alert, soldiers guard our bridges and airports. Our nation, "the land of the free and the home of the brave," lives with new restrictions and new gates.

Setha Low builds her book on a series of stories and interviews with residents, builders, architects, sales managers, and real estate agents. Much of it is impressionistic and descriptive, but it points to a major new area to probe. Are we resegregating America, after years of fighting to end segregation? Is this the new path our popular culture will take?

On May 20, 2003, the terrorism alert was raised to orange (high) because a number of indicators and warnings suggested a possible attack. Officials believed this was linked to al-Qaida, which had recently struck in Saudi Arabia and Morocco. Seventy-five people died, including eight Americans. Homeland Security Secretary Tom Ridge feared a wave of worldwide attacks that could include America. So up went the gates.

One main point emerges: "Don't Fence Me In," a once-popular song that told of our love of wide open spaces and unlimited freedom, is fast disappearing. Perhaps the new theme song will become "Stand Back and Stay Away from My Gate."

YES, BUT IS IT ART?

Can we find similarities between the art world, as it is narrowly conceived today, and the actual world in the twenty-first century, where national boundaries and traditions are giving way to globalization and a world linked by instant new technologies and concepts? I think we can.

In the art world disciplinary boundaries are giving way to innovative exchanges between art history and a wide range of disciplines. New links and ideas are championed and art history is taking on old connections and new meanings.

Hence the reemergence of such thinkers as Aristotle, Descartes, Kant, and Robert Hughes. Some would retrench and police the borders of where art history "belongs." Others choose another response: to engage debate precisely at the points where new boundaries are opening up and the disciplines are finally embracing new concepts— epistemological, aesthetic, and ethical. They are risking rethinking and rewriting art history. They are probing.

A 2003 exhibit at London's Tate Gallery, called "Museum at Work in the Age of Technological Display," is a case in point. Questions raised here include these: What is the relationship between art, the public, and the elite museum? What explains the almost universal enthusiasm from the public and critics when art is aimed at a mass audience?

The answers have to do with the exhibit itself, envisioned in the shell of a converted power station. When entering "The Turbine Hall," people find themselves in a new and incommensurable world. Art is more of an event than an object. It aims to "set forth the earth."

That idea was expanded, without that quote, years ago by America's pop artists such as Robert Rauschenberg, Stuart Davis, Jasper Johns, Roy Lichtenstein, Andy Warhol, and Claes Oldenburg, whose giant "Proposed Colossal Monument: Obelisk, Washington, DC— Scissors in Motion" dates from 1967, and whose forty-foot "Clothespin" loomed over central Philadelphia in 1976.

These, and other examples such as American experiments with op, pop, color field, and minimalism, were major contributions to what has come to be called "modern art." The idea of a fusion between outmoded art and politics has deep roots in America. By changing the language of art, we affect the modes of thought; by changing thought, we change life. The idea has long been present in American art, strongly stated with the "primitives" and the ashcan school. We err in giving most of the credit to Europeans, and filling our museums with their work, bought at astronomical prices.

The leading recent popular champion of American change, Andy Warhol, became famous for painting "200 Campbell's Soup Cans" in 1962 and for announcing that he wanted to be a machine. He loved the inert sameness of an infinite series of identical objects; the public loved what he did with soup cans, Coke bottles, dollar bills, Mona Lisa, and the head of Marilyn Monroe, silk-screened over and over again. These are examples of the merging of elite and popular art and culture, and of vast new opportunities.

Americans led the way once again. We are to take heart that America's art is once again crossing borders and oceans, opening up exciting new probes.

All this has alarmed the old art world, as Robert Hughes points out in *The Shock of the New* (1981). That world is still largely controlled by dealers who use gimmicks for prodding the market along by lending pseudohistorical weight to new art. That is changing. The words of some of the greatest artists ring out. "Museums are just a lot of lies," Pablo Picasso has said. "Work for life," Russia's Aleksandr Rodchenko proclaimed, "and not for palaces and museums." Times are changing fast: we must learn to change with them.

DREAMING AND PROBING

Sweet spring had a bitter taste in 2003. William Butler Yeats summed it up in a line written seventy-three years earlier: "Things fall apart; the center cannot hold."

Falling apart not only in what had been the Soviet Union, but also in the Middle East, Africa, Latin America, and, yes, in the United States. Consumer confidence fell, the rich got richer, the poor poorer. Petty politics plagued cities, states, and nations. When asked which presidential candidate they favored, half of Americans answered "none of the above." Had the American Dream, which fired up the world, finally melted down?

That dream first drew pioneers from Asia to a vast new continent when history was still blind. They crossed the frozen tundra and faced an endless winter because they dreamed of a better life.

Centuries later others crossed the uncharted Atlantic Ocean in tiny boats, drawn by the same dream, seeking a brave new world. In 1607, some reached what became Virginia (named for the virgin Queen Elizabeth). Michael Drayton called it "earth's only paradise," but many died of hunger, disease, and attacks. Yet the dream and the probing persisted. The struggling colonies grew and eventually won freedom. We have clung to it ever since.

The price of freedom is high. Our ancestors did not hesitate to pay the price, pledging their lives and sacred honor to win their independence. They agreed with Thomas Jefferson. They would rather die free men than live as slaves.

The contradiction of slavery had to be resolved. It was, in a Civil War that claimed over 600,000 lives. The American dream survived; slavery was abolished; and millions of immigrants, yearning to be free, came to America from all over the world. They are still coming. The nation's motto is *E Pluribus Unum:* Out of Many, One.

The many who came understood this. They learned new words, customs, laws. They were Americans now, breathing free air.

There was one proviso; *Pluribus* must always be balanced with *Unum.* The resulting tension makes democracy work. This is the American secret for success.

Have we forgotten the secret? Has the melting pot stopped melting? Are we becoming the not-so-United States? Have we lost our balance and the secret?

Democracies live with fragile alliances. A placid crowd can quickly become an angry mob. Violence can overcome reason. We end up substituting confrontation for compromise. In so doing we lose the longstanding price of freedom: to use it rationally.

Inequalities in gender, race, and ethnicity must be exposed and corrected; many are being addressed. But much remains to be done. We must strive for historically correct history, which is not the same thing as politically correct history. The very freedom our present dissenters and critics enjoy was made possible by discredited DWEMs (Dead White European Males). Supposedly it was Voltaire who wrote, "I disapprove of what you say, but I will defend to the death your right to say it."

To regard the early European roots of American culture as vile and wrong is short-sighted. We cannot change the past to suit today's special interest groups. Trashing valid past ideas and achievements will avail us nothing. Instead, we should broaden our outlook and cultural base without destroying them.

The great spokesman for black America, Martin Luther King Jr., was well schooled in Western thought and rhetoric. His speeches had a Ciceronian ring. His chief weapon was not a gun but a dream. We honor that dream today.

We have been in stormy and treacherous times, but we may be on the brink of a new renaissance. Suppose, like Sleeping Beauty in the fairy tale, we are waking from a long sleep, a deep cultural and economic depression?

That deep traumatic sleep might have begun with the assassination of John F. Kennedy, the Vietnam war, and the 1960s counterculture. Are we ready, finally, to greet a new century with a new, upbeat vision? Can we make the renaissance become a reality?

We live in perilous times. Will the terrorists strike again? Where and when? Is the Homeland safe? The airlines? Highways? City streets? Other nations around the globe? That world is badly divided by many issues and struggles with disease, genocide, and deep recession. What has happened to the spirit that might and should link us together?

A black American poet, Langston Hughes, urged us to hold fast to the American dream. Without the dream, hope dies. Many others implore us to deepen and widen our outlook. Are not dreams and probes related, linked together like Siamese twins? To profit from either, we must also deal with the other.

Dare to dream and to probe. Wise men and women have said this in all ages and places. Ezekiel, the great Hebrew prophet, did so many centuries ago: "Your sons and daughters shall prophecy, your old men shall dream dreams, your young men shall see visions."

Let our probes, dreams, and visions carry and guide us triumphantly into the new century.

"WE GOT HIM!"

No one expected it, but December 14, 2003, turned out to be a red-letter day for America. Before the day ended, the whole world knew why.

At 10:30 a.m. on December 13, U.S. troops got the final pieces that would end the longest man-hunt in history. By 6 p.m., Special Forces and six companies of infantry were on the scene, about ten miles south of Tikrit in the town of ad-Duar. For a while they found only a deserted shack full of debris, but at 8:26 p.m. an alert soldier saw a crack in the ground under a lean-to next to a mud hut. Underneath was a hidden door opening up to a styrofoam hatch covered by a rug.

Down the six-foot deep spider hole they found a disheveled man lying flat, with a fan for fresh air and a tube for a crude urinal. The man proved to be Saddam Hussein. The Lion of Babylon was a beaten man. "Don't shoot," he muttered in English and Arabic. He had grown a long beard and kept his tell-tale mustache, and admitted he was indeed their target. DNA samples confirmed this. Defeated and disgusted, the once-proud dictator had fallen.

It was given to L. Paul Bremer, Coalition Administrator in Iraq, to break the news, at about 3:50 p.m. Sunday afternoon:

"Ladies and gentlemen, we got him!"

Within hours the whole world was reacting with celebrations, dances, and celebratory gunfire. Saddam was hurried off to a secure place at the Baghdad Airport for interrogation. He seemed disoriented and defiant; but materials found in his spider hole provided valuable information about his network, money, and the people who were carrying on the guerrilla war. Asked why he had killed so many, Saddam called his victims "thieves and Iranian spies." He expressed no regrets or apologies.

President Bush's remarks were measured and restrained. He did not gloat, saying the tyrant had suffered a major setback. It was too early to know what impact the capture would have, or when and how he would be tried. But what a Christmas present for the Coalition and the armed forces! It was indeed a day of triumph. "High Value Target Number One" was a prisoner, found at the bottom of a six-foot hole near his hometown.

Paul Bremer ended his statement with this message of hope: "Now is the time for all Iraqis—Arabs and Kurds, Sunnis, Shias, Christians and Turkomen—to build a prosperous, democratic Iraq at peace with itself and with its neighbors."

Others will probe and ask what this meant for years to come. Some things were already clear. This was yet another emperor who had no clothes. He will never return to torture his foes or lead his small coterie back to their "glory days." The significance of this stark fact is enormous. No wonder this e-mail message circled the globe: "Happy Capture Day!"

OTHER VOICES

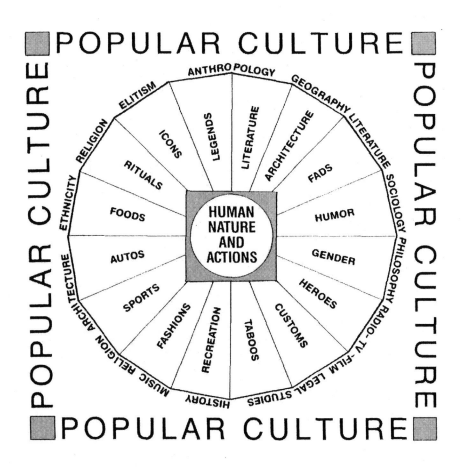

POPULAR CULTURE

Our three "Other Voices" in this section are all teachers, in a trouble-filled environment. Their students are far different from those they studied with or taught a generation ago. These two pictures, both taken at American colleges, reflect some of the new attitudes and cultures. Study the faces.

The Virus of Superficial Popular Culture Studies

Ray B. Browne

Popular culture studies have become a large part of academia and everyday assumptions and analysis. Thus it is imperative in order to avoid corruption and misunderstanding that everyone have full knowledge and comprehension of what the subject is. Yet understanding has fallen into the trap of wrong or incorrect definitions. As Kenneth Burke once observed, "Every way of seeing is a way of not seeing." Therefore we must all at least understand the basics of popular culture definitions.

By the term "popular culture," we mean all aspects of the culture we inhabit: "the way of life we inherit, practice, and pass on to our descendants; what we do while we are awake, the dreams we dream while asleep. It is the everyday world around us: the mass media, entertainments, diversions, heroes, icons, rituals, psychology, religion —our total life picture."[1]

Misunderstandings and misuses of the term *popular culture,* one somewhat unintentional and the other willful, have dragged us into a wolf trap from which is emerging disservices that are spreading like a virus, weakening the justice paid to the subject and to the imperative of deep understanding and analysis. In our cyber age of specialization we need some cohesive force to hold society together and to make sense of it for us. We need to know how to locate our position in the human and physical universe, in other words, to grasp a firmer hold on *reality* rather than on *virtual reality,* and that force as the democratizing power in society. Failure to understand the differences and the resulting forces might result in what lawyer Richard K. Sherwin fears as "increasing distortions within the discourse competencies of both the public and the judiciary" when, as he says, himself confusing the terms, "the law goes *pop*" (p. 242).

Perhaps a reminder of the background of the subject might be useful. Throughout the history of western Europe from the fifteenth century on and in American society from the beginning, the elite elements in society, in order to separate themselves from the less fortunate have looked upon the larger segments in society as less deserving and capable and their customs and ways of life as inferior. Therefore the elite ways—their cultures—were generally regarded as superior. These differences, once defined, had to be maintained by constant iteration and demonstrations. The upper classes suppressed the lower. The lower, though grumbling all the time, found means and avenues to rise above their designated status, especially in such new and democratically developing countries as America. Through the years they have managed to live in a separate but somewhat equal society in which all cultures are roughly the same, since all forms of expression are inseparably bound to culture, either leading, following, or enjoying presentness as guaranteed by the ballot and resulting laws.

Popular culture has not received the understanding and dignity it deserves because of academic carelessness and penchant for vernacular shorthand. In 1970 the first and trailblazing study of popular culture, Russel B. Nye's *The Unembarrassed Muse: The Popular Arts in America,* began to legitimize popular culture study in academia. The key words in this book were, of course, "popular arts," but most readers misread them as "popular culture" or the arts as the only kind of popular culture. This misreading has through the years been strengthened by a verbal laziness and built-in construction of the word *popular* that invited abbreviations to *pop,* which came to mean cheapness or tawdriness in society in general and the everyday entertainment world—like movies, music, radio, and sports—in particular.

Because we are the strength or weakness of our language "pop culture" became a term of superficiality or opprobrium. Many scholars who approved the study of the subject used the name—either in its dictionary form or its colloquial form—with a blush or admitted apology. It was a kind of closet interest. When the *Journal of Popular Culture,* the first publication dedicated to the study of popular culture, was begun in 1967, for example, many reputable scholars reached into desk drawers and pulled out serious studies on popular culture studies. There are even a few PhD dissertations gathering dust in drawers. Immediately the elite—especially those that felt threatened by the newly opened door to new fields of human society—started condemning the field both as "popular culture" and as "pop," a sharp criticism that has continued to this day.

The term spilled over into the serious study of everyday culture, infected it and continues to result in a spreading virus that weakens the democratic body politic. One of the powerful forces is the condescending force among the scholars who approve of the discipline but continue to sue the denigrating terminology in classes of popular culture by both the students and instructors, both undergraduate and graduate. Recently, for example, a university professor announced on e-mail a graduate culture course in "Pop Culture."

This perhaps unconscious though damaging derogation extends to the authors of books on the subject. For example, David Jackson (2002) in *Entertainment and Politics: The Influence of Pop Culture on Young Adult Political Socialization* states that young people are "currently being influenced by popular culture" (p. 9). But he fails to notice that young people are consumers—critical or otherwise—of all kinds of popular culture. Calling it "pop" culture, instead of *popular* culture as many commentators do, instead of complimenting the democratic way of life, is really insulting it.

The full development of popular culture understanding and studies is held back by another powerful force that shines in the sunlight of underdevelopment. The bright light of the glitter of popular culture studies might be called the Tut Factor, after the glowing wealth found in the tomb of the young pharaoh. Among scholars this glitter might be overvaluation of a Citation Index, to which no Discrimination Index is attached, as the litmus test of value. In other words, use alone validates the correctness and value of aspects of culture studies. Dependence on thoughtless, traditionally accepted knowledge is dangerous because it is only compounded when it is restricted to a narrow range of uninformed or misinformed knowledge. As has been voiced by Plutarch, and numerous others: "Custom is almost a second nature."

Just as it is difficult to break the restraints of tradition in America it seems even more painful to leap the overseas boundaries that surround our cultures. Academically and culturally the Atlantic has not been bridged as successfully as has the Pacific. Europeans, for their own reasons, do not read American scholarship—or extend their reading to the proclaimed highlights—and Americans do not read

European. Sometimes it is a matter of translation or availability, but more often it seems to be reliance on starred and cited references. Despite the cry for years about internationalizing scholarship, most of us, for one reason or another, do not investigate scholarship that is not prominently displayed even though we froth at the mouth when advancing arguments built on information that was incomplete or incorrect when made available and is still doing more harm than good as it becomes more and more a relic of history.[2]

One of the worst abusers of the Tut Factor is the media. Reporters for newspapers and the electronic media deal in the daily power brokers and actions of life. Today most newspapers have what they call a popular culture reporter who has been directed to limit himself or herself to entertainment—movies, television, popular individuals, best-sellers, etc. Other items of culture—which constitute most of life—are left to other reporters. In the eyes of these popular culture reporters, entertainment becomes "pop culture" and information covered by other reporters becomes either "pop culture" or just news.

Book publishers—both academic and commercial—are doing the study of popular culture a great positive service by jumping into the melee of publishing popular books on all subjects, and in so doing placing a stamp of societal—or at least commercial—approval on their products. In their haste to get a book published that will sell well they sometimes muddy the waters of clear understanding. Regardless of the subject, writers of flap copy frequently outline the "pop culture" elements in the book though it has nothing to do with the entertainment media.

At times also the book fails in its endeavor to be of real consequence. Manuscripts for publication are sent outside for peer review. These reviewers frequently have only the scantiest knowledge of the subject and virtually none of what popular culture is. Therefore the reviewers read the manuscript for its old qualities but not for its insights into democratic cultures. As a result the publishers bring out superficial or shortsighted books that do not do justice to the depth and breadth of its cultural subjects but nudge the field of popular culture studies forward somewhat.

Another positive force driving forward the study of popular culture is the Popular Culture Association/American Culture Association with their annual international conference and its ten annual regional meetings that open the doors to all subjects of interest to anyone. The several meetings of the Modern Language Association and the yearly meeting of the American Historical Association, the Organization of American Organizations, and the numerous other scholarly organizations in the humanities and social sciences frequently make presentations on everyday subjects. One organization—American Studies Association—and its publication, *American Quarterly,* resist including the study of popular culture in their agenda. In *The Futures of American Studies,*[3] containing twenty-three essays, the editors attempt to outline what is in store for American studies in the coming decades. The book is built on the paradigm outlined more than two decades ago by Gene Wise (1979), "Paradigm Dramas in American Studies: A Cultural and Institutional History of the Movement." But the editors stumble before getting to the end of Wise's essay. He outlines very clearly in the essay that he thinks popular culture is "one of the more energetic forces in American academic life" (p. 316) and it has continued to stimulate the broad field of the humanities and social sciences ever since his essay was published.

Nonacademics in all fields, recognizing the importance of stepping outside their conventional boundaries, are turning over new leaves. Historians, museologists, art critics, architects—everyone is looking for new knowledge and stimulation. Historians are probing the growing field of historical fiction to feed on the knowledge and imagination of history crime writers. Historian Robin Winks recently developed a new respect for historical crime fiction because "Historians moved too far away from their origins, as storytellers; now storytellers may bring historians back those roots, to the benefit of both ways of exploring the past" (2000, p. x). One historian even teaches an introductory course in history through a half dozen historical crime novels.

Other professionals, especially those who walk in the popular culture all the time, are searching around them to read what is written everywhere. Politicians and lawyers must know whereof they speak otherwise they offend catastrophically. In 1999 I edited a book called *Laws of Our Fathers: The Constitution and Popular Culture,* in which a dozen authors demonstrated how the U.S. Constitution, and especially the Bill of Rights, has been directed and shaped by the everyday culture it heads. Among recent studies of the law in our culture, other legal experts recognize the interworking between popular culture—though many are primarily concerned with the media—as one of the most powerful social forces and law.

Scholars of the distant past, and scientific fields of the present—paleontologists, archaeologists, Egyptologists, and other such scientists—increasingly are giving up the Tut Factor in order to understand the past and interpret it for our knowledge and understanding. It is a matter of maturity. Egyptologists have turned to the workers who built the pyramids, while archaeologists of Mayan culture are still looking for the gold artifacts. Increasingly, however, they will discover and appreciate that the value of the past also includes the unshiny as well as the glittery. Not all that glitters is gold and not all that does not shine is worthless.

Increasingly academicians—and the public—are discovering that in a democracy—or a society hungering for democracy—the proper study of democracy is the culture of democracy, and are turning to everyday cultures for appreciation of the world in which we live. The cultures are actually encompassed in the proper definition of the humanities. This was defined long ago by Harvard psychologist Robert Coles as the elements of humanity which defines us as human beings. He pointed out that they belong not to a special class but to us all, the humble and the powerful, the knowledgeable and uninformed.

To get this message across and the mission accomplished a modified, new method of communication must be realized—a new language, as it were, or a universal bond to protect society from monopolistic and power-controlling media communication. Language has always been an elusive skill of communication that has been controlled by individuals or groups for special purposes in addition to passing on messages from one individual to another. Words and rhetoric are labels to be attached to items and phenomena. As such they have had two existences, static and alive. They are like identifying labels on packages or boxes. They can remain mere labels or they can identify forces underneath. Kept quiet as mere expressions, they only identify. But all—even the mildest—can take on a power of their own, bursting into a new thrust when vocalized heatedly.

They take on much of this thrust—though often so quietly they are almost misunderstood—when used generically. Words and phrases follow needs and take on this generic strength when used so often in society that their original edges of distinction are worn away until they become general vocabulary. Three such examples are "paper clip," "Coke," and "McDonald's," terms loaded with meaning of approval or disapproval.

The paper clip is looked upon by one commentator, Owen Edwards (2002), as the finest achievement of mankind:

> If all that survives of our fatally flawed civilization is the humble paper clip, archaeologists from some galaxy far, far away may give us more credit than we deserve. In our vast catalog of material innovation, no more perfectly conceived object exists.[4]

Although the Gem paper clip nears perfection, great minds are still challenged to further develop it so that it achieves its purpose of properly holding pages of paper together without puncturing the paper. The term *paper clip* is then a perfect word in works in progress.

"Coke" and "McDonald's" are works of love or hate depending on who is using them and who the audience is. "Coke" is the term used for all soft drinks, and will perk you up or rot your teeth, depending on your point of view. "McDonald's" is one of the most explosive generic terms now in the English language, especially in the vocabulary of food, fat, and failing national health. Societies have always had the need to grab a bite of food from someone on the street as they rushed off to business, pleasure, or war. Little care was paid either to cleanliness and healthiness; food was the desire. Today, it seems that every street corner and commercial strip is anchored by fast-food restaurants such as McDonald's, Burger King, Pizza Hut, Wendy's, or any of another two dozen chains. All are accused of serving fat-laden fried potatoes and hamburgers. At the moment Wendy's is enjoying the broadest smile because of its alternative menu, but undoubtedly that chain will come under fire when and if the others are subdued. Words individually, as symbols and as generic terms, take on the life of the user.

Words and rhetoric rise like a pyramid from the surrounding terrain. The greatest changes and most significant power are generated most radically during times of military conflict and social upheaval. Individual rights are more freely granted and society changes, as capitalistic democratic inventions and unregulated marketing skills are developed to distribute the new changes in society. The bottom tier of the pyramid consists of everyday speech, which at its simplest communicates effectively. Much more effective are those words charged with power-driven emotion, as in Julia Ward Howe's Civil War triumphal song "Battle Hymn of the Republic." The first two lines were powerful enough to alert and frighten a nation: "Mine eyes have seen the glory of the coming of the Lord / He is trampling out the vintage where the grapes of wrath are stored."

Sometimes, given the opportuneness of the occasion, a song—with the most effective mixture of words and music—will overwhelm a nation. Such was the fate of the Civil War song "Dixie." This song was written in 1859 by Daniel Decatur Emmett, an Ohioan, as a "Walkaround" to end a northern minstrel presentation. The apt words and lively tune immediately caught the fancy of all listeners and was powerful enough during the Civil War both to divide and unite the Union and the Confederacy. It was Lincoln's favorite song. It was President Davis's favorite, also, played at his inauguration as President of the Confederacy and at his resignation of that office. Its words were so powerful in the Confederacy that the song sported more than 200 versions.

Words and rhetoric are powerful and lasting when spoken. Lincoln's Gettysburg Address was 250 words delivered in less than two minutes. With its hope that "government of the people, by the people, and for the people, shall not perish from this earth," it is undoubtedly the most quoted passage in the English language outside of passages from the Bible.

At the top of the pyramid, perhaps, words become profanity, or are used for profanity, and take on a violence of their own. As collective violence, it is war or an individual or group substitute for war. "During war, the effect of violence upon language is amplified and clarified," says James Dawes. "Language is censored, encrypted, and euphemized; imperatives replace dialogue, and nations communicate their intentions most dramatically through the use of injury rather than symbol."[5] As Dawes implies, language, like form in the architectural context, follows use. It is shaped, choked, or vitalized as necessity dictates.

Sometimes, when language is used merely for communication, particular persons, Shakespeare for example, seem to set the pattern. Famed British actor Christopher Lee, veteran of 250 movies and star of the J. R. R. Tolkien trilogy *The Lord of the Rings,* thinks Tolkien is "the greatest inventor of language and legend"[6] in his eighty-year lifetime. Abraham Lincoln, as discussed previously, was very much aware of the vagaries of meaning in words. "We all declare for liberty," he said during the Civil War, "but in using the same *word* we do not all mean the same *thing*."

To paraphrase the nineteenth-century poet Alfred Lord Tennyson, "words, like Nature, half reveal / and half conceal the Soul within" ("In Memoriam, Prologue," Part V, Stanza 1). But it is the hidden half we must beware. The drive to create and control language—and thus the thinking—prevails today among propagandists of all flavors and degrees of self-servers: advertisers, spin doctors in all trades and vocations, and inward-looking tribal and consanguine groups.

The urgency is for society to free itself from the shackles of these wordmasters, especially those in politics and the ever-expanding reaches of law and lawyers. Richard K. Sherwin, a lawyer himself, emphasizes the need:

> New rhetorical handbooks must be written, new media laboratories must be created, to reflect current conventions of meaning-making and to bring to bear new insights regarding how legal meanings are constructed, altered, or suppressed. Unconscious and symbolic forms of expression as well as popular cultural and cognitive constructs need to be included within a new interdisciplinary domain of law, media and cultural studies.[7]

One could hardly find a more convincing statement urging the understanding of popular culture and its intertwining with "high" culture. The two are both strands of the same cultural rope.

Both Morse and Sherwin are describing some of the forces of cultural suasion and ways of protecting society from them. These are descriptions of power and avenues of movement. All of us should be aware of them and alert to their impact. But before we start moving them perhaps we should understand why on a large scale the need is pressing. On a cultural level we are in the midst of a powerful ocean swell which is changing our vocabulary and rhetorical outlook, what might be called a near-irresistible virus that is weakening us and directing us down the path to cultural inexactitude and confusion. In their new book *Enviropop: Studies in Environmental Rhetoric and Popular Culture,* editors Meisler and Japp set out to demonstrate "how popular culture has rhetorically constructed environmental issues" (p. 1). They then narrowly limit their definition of popular culture to those forms of mass media, games, food, music, shopping, and other daily processes of activities and especially movies and newspapers. But the editors twist their definition with the caution that "the materialism promoted by popular culture encourages acquisition, leisure and recreation" (pp. 8-9). Again we are on the slippery slope recognized by Lincoln that we all use the same *words* but with different *meanings.*[8]

In a rhetorical culture where anything means everything and anything, only confusion results, a state of bewilderment clearly demonstrated in Lewis Carroll's *Through the Looking Glass* in a conversation between Humpty Dumpty and Alice:

> "But 'glory' doesn't mean a nice knock-down argument," Alice objected.
>
> "When *I* use the word," Humpty Dumpty said, in rather a scornful tone, "it means just what I choose it to mean—neither more nor less."
>
> "The question is," said Alice, "whether you *can* make words mean so many different things."
>
> "The question is," said Humpty Dumpty, "which is to be master—that's all." (Carroll, Chapter 6)

In the question of the spreading virus of confusion there may be no immediate mortalities at stake. It would be nice to have people know what they are talking about before they start acting on a subject as broad, proud, and important as democratic culture.

NOTES

1. Ray B. Browne, "Popular Culture As the New Humanities," in Ray B. Browne and Marshall Fishwick, *Symbiosis: Popular Culture of Other Fields.* Bowling Green, OH: Popular Press, 1988, p. 1.

2. The British, perhaps surprisingly, are more careful in their terminology than are Americans. In his latest book, for example, *Bestsellers: Popular Fiction Since 1900* (New York: Palgrave, 2002), author Clive Bloom does not use the term "pop" culture, and when referring to the media he calls them movies, television, popular music, etc. Canadians, likewise, are more likely to be precise. In the entry on popular culture in the new *Encyclopedia of Literature in Canada,* edited by H. W. New (Toronto: University of Toronto Press, 2002), the author of the popular culture entry does not use the term "pop" at all.

3. Donald E. Pease and Robyn Wiegman, *The Futures of American Studies* (Durham: Duke University Press, 2002).

4. Owen Edwards, *Elegant Solutions: Quintessential Technology for a User-Friendly World.* Quoted in Henry Petroski, *The Evolution of Useful Things* (New York: Vintage, 2002), p. 63.

5. Dawes, James. *The Language of War: Literature and Culture in the U.S. from the Beginning Through World War II.* Cambridge: Harvard University Press, 2002.

6. Quoted in Jennifer Dorm, "Christopher Lee, the Wizard of *The Lord of the Rings,*" *British Heritage,* 24(1), January 2003, p. 40.

7. Richard K. Sherwin, *When Law Goes Pop: The Vanishing Line Between Law and Popular Culture* (Chicago: University of Chicago Press, 2002), pp. 240, 246.

8. Mark Meister and Phyllis M. Japp, *Enviropop: Studies in Environmental Rhetorical and Popular Culture* (Westport, CT: Prager, 2002).

BIBLIOGRAPHY

Browne, Ray B. and Glenn Browne, *Laws of Our Fathers: The Constitution and Popular Culture,* Bowling Green, OH: Popular Press, 1999.

Browne, Ray B. and Marshall Fishwick, *Symbiosis: Popular Culture & Other Fields,* Bowling Green, OH: Popular Press, 1988.

Browne, Ray B. and Lawrence A. Kreiser Jr., *The Detective As Historian: History and Art in Historical Crime Fiction,* Bowling Green, OH: Popular Press, 2000, Preface, p. x.

Brooks, Kevin, Kathleen Blake Yancey, and Mark Zachry, Developing Programs in the Corporate University: New Models. *Profession 2002.* New York: Modern Language Association of America, 2002.

Dawes, James. *The Language of War: Literature and Culture in the U.S. from the Beginning Through World War II.* Cambridge: Harvard University Press, 2002.

Dorm, Jennifer. "Christopher Lee, the Wizard of *The Lord of the Rings,*" *British Heritage* 24(1), January 2003, p. 40.

Edwards, Owen. *Elegant Solutions: Quintessential Technology for a User-Friendly World.* Quoted in Henry Petroski, *The Evolution of Useful Things.* New York: Vintage, 2002, p. 63.

Jackson, David. *Entertainment and Politics: The Influence of Pop Culture on Young Adult Political Socialization.* New York: Peter Lang, 2002.

Meister, Mark and Phyllis M. Japp, *Enviropop: Studies in Environmental Rhetorical and Popular Culture.* Wesport, CT: Prager, 2002.

Nye, Russel B. *The Unembarrassed Muse: The Popular Arts in America.* New York: Dial Press, 1970.

Pease, Donald E. and Robyn Wiegman. *The Futures of American Studies.* Durham: Duke University Press, 2002.

Plutarch, *Rules for the Preservation of Health,* 18.

Sherwin, Richard K. *When Law Goes Pop: The Vanishing Line Between Law and Popular Culture.* Chicago: University of Chicago Press, 2002, pp. 240-241.

Winks, Robin W. "Preface," in Browne, Ray B. and Lawrence A. Kreiser, *The Detective As Historian: History and Art in Historical Crime Fiction.* Bowling Green, OH: Popular Press, 2000, p. x.

Wise, Gene. Paradigm Dramas in American Studies: A Cultural and Institutional History of the Movement. *American Quarterly,* 31(3), 1979, pp. 293-337.

Teachers, Teens, and Technology

Katherine Lynde

It's three years into the new millennium. I communicate through e-mail, my husband lives through his Palm Pilot, and my kids watch the DVD player while traveling in our new mini-van. It's the age of technology. However, I'm a middle-aged techno-wannabe who grew up in the heyday of Star Trek sci-fi, in which futuristic oohs and aahs have become the toys and tools of today. When I was a teenager, there was no such thing as the personal computer. I vividly remember when a mainframe computer required an entire room of its own. Let's not forget that during my lifetime, most homes had black-and-white television sets, and sitcoms reflected the best of American values. Where does that leave me and many of my peers? Behind!

The public school system is perhaps the best place to find evidence of technological disparity between many adults and the next generation. As an English teacher, I have spent many hours advising students to avoid using clichés to make a point, but here the cart-before-the-horse-scenario works well. Students are submitting innovative technological products in response to assignments and implementing available software most of my colleagues have never even heard of before. Not to mislead my reader, several teachers are quite savvy with contemporary programming and software, but they definitely comprise a frightening minority.

Two years ago, my school system switched to computerized grading, and the monster—fear of change—raised its ugly head. Technophobes and techno-wizards divided themselves, with Integrade software as the boundary. Some teachers were merely resistant, but soon found themselves relieved from the drudgery of paper, pen, and calculator grade averaging. Some are *still* grumbling. Eventually, we will be required to learn computerized attendance record keeping. For those teachers who are still grumbling about computerized grading, that may be the last straw.

All of this may seem ironic, considering that my school system resides in the heart of what has been dubbed as "the most-hooked-up town in the United States—the electronic village." It has been especially infuriating since I attended a university that reinforced technological training as a necessity. Hypertext—now "old school" only six years later—was commonplace as a means of composition. Until this school year, the high school housed only one computer lab, but it was also used as a classroom, which means it was only available part of the school day. The library also maintained a bank of computers. Between the two facilities, less than fifty computers were available at any one time for approximately 1,200 students. Do the math.

This year, two new computer labs opened and the library received new computers. The math gets a little better. During my school duty period, I act as a computer lab monitor. It is rewarding to be present when progressive, unafraid teachers bring classes in to explore research options, to compose assignments, and to use exciting software. Students with different levels of ability and with different levels of socioeconomic status all have the opportunity to share the same benefits of the Internet, Spellcheck, and Grammarcheck. For the students in my high school, technology is actually erasing the disparity between the "haves" and "the have-nots."

In the early years of my high school teaching experience, I readily noticed that the socioeconomical polar extremes differentiated the classroom academic levels. Middle to upper-middle-class students dominated the highest-level courses, and the lower-middle to lower-class students were contained in the lower-level courses. Admittedly, many factors determine motivational student interest, but the presence of home computers is a powerful one. Since I teach both high-level and low-level English classes, I have witnessed an amazing shift—most students now have computers at home. This may mean that the lines will become blurred between the knowledge base of the social classes—an interesting dilemma for the sociological hierarchy's status quo and for my school's leveling policy.

Of course, computer technology presents dilemmas of its own—primarily, cheating. Teachers idealistically assign homework and essays that challenge students' skills and assumptions. Students realistically begin the assignment by accessing the Internet and—blink—hundreds, thousands, or tens of thousands of cites, excerpts, and essays for sale beckon at their fingertips. It's the latest newsworthy facet of education. For example, the University of Virginia, a well-respected university, caught and prosecuted students for plagiarism and made national headlines for doing so. Plagiarism has become commonplace. My colleagues constantly catch students red-handed. It is difficult to teach teenagers that kidnapping another person's work is wrong when it seems so endorsed by a recommended source: the World Wide Web. In the English Department, teachers often discuss ways of circumventing plagiarism. Limiting student composition to in-class, impromptu, handwritten essays seems like an obvious solution; however, it also is limiting in teaching the writing process. In other words, we cut off our noses to spite our faces. Teachers force students to use the writing process—planning, researching, drafting, revising—by cutting the process to a rough draft.

For the students who do choose to complete their own assignments and to write their own papers, another temptation looms to threaten their imaginations: Instant Messaging. Parents may not know that their children are communicating with anyone else while researching and typing on the computer, but too often, teenagers are involved in an ongoing conversation with a friend while trying to maintain higher cognitive skills. Educators spend extra time assessing papers that are not cohesive, seem to wander, and have no focus—imagine that for a moment. IM'ing *me* might be helpful. Then we could call it Instant Collaboration. Students could use each other as collaborators as well, offering suggestions and advice.

Of course, conversing from a keyboard has its advantages. Most Americans are familiar with the dangers of chat rooms and their counterparts—the dangers of anonymous transmissions. However, in a classroom or on an e-mail mailing list, keyboard communication can be a fantastic enhancement. Online discussions allow students who normally do not contribute in an open-discussion forum to have a "voice." On the other hand, students who dominate class discussion must take turns adding to the conversation. Computers can be the great equalizers. Unfortunately, online discussion is not as prevalent in secondary education as it is at the university level. Our school's Web site has a link for teachers to create Web pages with content and assignment information. Some teachers have pages available, and some may have students set up on a mailing list; it would be even better if more would use software programs for classroom chats. But couldn't we lose something with computerized communication and learning?

Teachers who wish time would march backward recall the day when students sat at the sandals of the great educators like Socrates, absorbing wisdom and asking the important questions. Is that concept humming along toward oblivion? A few of my students have been enrolled in online education, taking courses through various colleges and universities. While I acknowledge that computer communication is effective, I question whether it can be a replacement. Advancements in audio/visual capability make it appear as if students and teachers are actually together. To people who attended old-fashioned schools and who teach in an interactive, we-*are*-in-the-same-room classroom, it seems too clinical somehow. I ask, "WWST?" (What would Socrates think?)

It was sensory input that initially drew me to English and to books. Words on a physical page are incredible. Words on pages placed together between covers are even more incredible. Books have texture, books have substance, and books have scents. No—that's not a misprint. I meant scents, not sense. I have always been a closet book-smeller. In the past few years, I've discovered that I'm not alone; there are a lot of us out there. Computers have screens. My students are very accustomed to reading from screens. I'm doing it right now. But a dog-eared, margin-filled, musty-smelling book is a companion. During a graduate school seminar, we explored the possibility that online texts would replace the need for books. There were advocates in both camps. I'm not positive that it would be a great sense of loss for today's teenagers, but I believe it would be.

Most educators, including myself, must decide whether any real loss will be suffered by accepting advanced technology or if we are simply hanging on and embracing that old monster. One certainty exists—students are moving on with or without us. Understand that teachers feel overwhelmed daily with planning, researching, designing, and grading; however, they must make room for technology training. My school system provides training here and there, but it needs frequency and extensiveness if educators are to keep up with the students they are charged with educating. Not only will the cart leave the horse far behind, but it will find it doesn't need the old nag at all. Is that such a bad thing? WWST?

I don't have any concrete answers; I merely bring the situation to light. I might even be part of the problem. I'm still wondering what happened to writing a good letter, when my husband's own memory disappeared, and why my children can't look out the van window. But I do know one thing about the next generation's education—secondary teachers must either stand the heat or get out of the kitchen.

The Realm of Splogia:
A Report to the World Anthropological Legation

James Combs

The Realm of Splogia is of interest to anthropologists as the fore-most society committed to the innovative cultural proposition of *sploge*. The culturologic of sploge is not unknown in history, since there has been much spread and spoilage, spillage and sprawl, splurge and satiation, ugliness and ruin, waste and abandonment, and so forth. It is significant that archaeologists dig up and study human leavings, since our residue—what we leave behind—is often our only legacy. But the Splogians appear to be the first people for whom sploge is the primal myth of their culture. Virtually every culture, in-cluding disreputable groups such as pirates and the Mafia, featured in their mores and folkways a principle of conservation: things of importance were to be saved, used wisely, taken care of, and passed on. This included property, public edifices, natural resources, human relations such as family and friends and community, and both private and social inheritance. Even though many cultures allow or even mandate occasions of indulgence and revelry, such as carnivals and holidays, their return to normal life mandates a certain degree of re-spect for and practice of conservative principles. It is common for moral codes in a wide variety of cultural traditions to embrace self-control and parsimony and condemn self-indulgence and willful waste. A culture may permit the use of alcohol, but condemn drunk-enness and certainly alcoholism; similarly, various peoples may value the creation of wealth, but find appalling wastrels and misers, two ex-tremes of misusing the gift of fortune. Cultural conservatism, with some notable exceptions, dictates norms of restraint and obligation that discourage and even penalized excessive practices in violation of community decency and probity. This common principle applies to nomads and hunter-gatherers and traders and craft and industrial peoples alike.

The Splogians seem to have pioneered a new form of life that abandons the principle of conservation in favor of the principle of *sploge*. Rather than conserving important things, the new culture is committed to splogation, the immediate use and even abuse of things until they are no longer of any use and thence neglected or discarded. The principle of splogation rejects claims of moral absolutism and relativism in human affairs, maintaining there are no absolute principles (e.g., "Respect nature," "Treat others fairly") nor relative merits ("Respect plural perspectives," "Consider alternative choices"). Rather Splogians proceed on the basis of a rude moral opportunism: What can I gain from doing this without regard for archaic principles of conservation and considerations of dire consequences? The Splogian mind focuses on the immediate, with contempt for the lessons of the past and disregard for future generations. They act with confident moral clarity. Rather than knowing one big absolute thing, or knowing many related things, they know nothing, or at least nothing that interferes with immediate interests leading to gain. To the extent the Splogians profess a common faith, it is based on the core value that it is more blessed to receive than to give.

The Splogians occupy the temperate zone of North America at the present, having slowly but surely superceded the people of a prior culture known as Localia. The Localians occupied the land for a long period after the defeat of native aborigines, but shared many features with their predecessors. As their tribal name implies, they occupied many local areas and tended to live and die a local life. Life in locales was centered in towns surrounded by country agricultural and pastoral enterprises, and in the cities centered in neighborhoods often with ethnic identities or mixes. In most locales, whether rural or urban, the Localians possessed many survival skills now lost. They could sew and knit and weave and mend; butcher and dress meat and can and preserve and cook; fix or repair a wide variety of items, from plumbing and wiring to cars and appliances; save money and avoid debt; live in a local area close to family and friends most of their lives; and pass on property and wealth and memory to their offspring. By contrast, their successors on the land, the Splogians, are a more restless and rootless people, whose ambitions for More have soared with their accumulation of wealth and the technological means to sploge the world. The Localians tended to think that the country was the rural place for farms and pastures which supplied the towns and cities with food and raw material, and that urban places supplied the country with goods and services. There was a great deal of continuity to Localian life, and relatively limited expectations. Although the Localians had their social and intellectual limitations, their passage as the primary form of life is North America has become the object of much nostalgic memory and depiction.

The Localians were eventually displaced by Splogian society, which observes the memory of their predecessors but eschews any of their practices. Many things that appear to be continuous in fact are not. Religion was important to the Localians, but less so to the Splogians, who use religion as a justification for their exploitative and expansive practices. Splogian religion does not preach simplicity or restraint or altruism; rather it supports the principle of sploge through a gospel of wealth and prosperity. It is true that Localians did a bit of sploging in the old days, but Splogia has committed to splogation on a scale hitherto undreamed of in the Localian mind, without any local institutions such as churches or even mere communal considerations—getting along with your neighbors—to restrain them. Virtually everyone now seems complicit in the social project of splogation, since sploge is after all the distinctive and almost unique feature of this new society. So quite understandably the Splogians point with pride toward any new and successful project of splogation—for instance, supermalls covering vast acreages of previously fertile agricultural or pastoral areas.

Splogia is ruled by a stable and self-perpetuating elite caste called the Moronians. The Moronians are distinguished by their desire to banish all critical thought from their realm in order to pursue the maximization of splogation, and by their avowed contempt for intelligence. As an elite caste, they have succeeded in their effort to not think about things that they don't want to think about, and only think about those things they want to think about. To think as a Moronian requires not only an unreflective commitment to splogation, but also mastery of the confident attitude of *No Doubt*. This attitude is expressed in a series of affirming truisms, such as "Even too much is not enough," "Bigger is better and biggest is best," and quoting the Roman emperor Nero addressing his newly appointed magistrates charged with tax gathering, "You and I must see that nobody is left with anything."

The Moronians loudly profess adherence to the tribal gods inherited from the Localians, but in fact they worship the god Splogeh, a more refined and ambitious version of Mammon. Splogeh is the god of acquisition and accumulation, and is often reverentially cited by the Moronians for the commandment, "Thou shalt get all that thou can." The temples of Splogeh can be found everywhere, in solemn places of worship called "banks," wherein sacred objects named "money" are kept as a sacred offering to the god. (The central temple of the state religion is called The Federal Resplurge, found in the capital city of Splogia, Splogington.) To the extent that one accumulates money, one is more likely to be counted among the Elect favored by Splogeh. But Splogeh looks with greatest approval upon those who use money to spread the gospel of sploge across the land. Money is made into the votive offering known as "venture capital," and creates a corps of disciples called "developers," who go forth to spread the Word of Splogeh and build temples in his honor for worshippers to come and offer money as tithe and offering in seeking Splogeh's favor. (Splogeh's wrath is especially loosed on those who save money, since one of the god's primary commandments is, "Splurge!") These good and faithful servants often join religious orders separated from the world in "gated communities," so they may witness and share their joy in Splogeh's reward. While inside, they attempt to perfect their spiritual life through the practice of the discipline of *No Thought* and the mantra of *No Doubt.* (They often chant sacred truths, such as "Reward is its own virtue.") As true believers, the Gated People are thought to be the chosen of the Earth, since they have dedicated themselves to the service of Splogeh through the sacred projects of development.

For the anthropologist, however, this project is a fanaticism stemming from the pathological condition known as *developmentia.* The ancient myth of the Localians they called "progress" has metamorphed into the Splogian myth of "development." Whereas progress was moderated by community considerations and local habits, development knows no impediments. Observers often have noted that development often violates local life by intruding on tradition and habitation. It is no wonder that powerless groups such as the Localians regard development as a disease spreading across town and country, since the march of development seems relentless and often pointless. But for the Moronian caste, development is sanctified work and a finished development—a mall, a housing project, a skyscraper, a road, an industrial park—a noble and beautiful improvement on the natural or old. Moronian developers often cite the sacred text of Splogeh, "Sprawl is good." (Sprawl is the conversion of places of natural beauty or old charm into places of constructed beauty and new appeal, such as a mall covering a former forestland.) In their own defense, the Moronians often cite the injunction of holy writ: "Thou shalt spread sprawl everywhere." Moronians often point with pride to their ability at despoilation: A meadow is nothing unless it is full of McMansions; a cove is a bore without condos; a wooded trail is better if it is asphalted and accessible by cars; rivers are just running water unless they are full of recreational boats; and mountains are only hills unless they are topped by discount malls. Sprawl is sponsored by that Moronian mystical body, the "corporation," which is the nobody that does everything even though it is actually nowhere. This ethereal entity does exist through the Moronian practice of "owning." As a disciple of Splogeh said, "To own is to be." The mystical corporation may be noncorporeal, but since it owns, it exists. Paradoxically, the Localians are real bodies that live in real communities, but since they do not own, they do not exist. The Moronian corpus called "the law" codifies this mystery and enforces its decrees.

The Moronian state is headed by ornamental figures who are publicly displayed in order to inspire confidence and elicit deference from the lower orders. Even though the Moronian elite are a self-perpetuating caste and the system that sustains them rigged in their favor, they enunciate platitudes of benevolence and opportunity. Although unbridled developmentia may strike those outside the pale as insane, it is defended as an act of heroism that advances the quick march to perfection. For the Moronians are idealists, who dream of a perfect world which is completely developed and covered with sprawl. At this point, the Moronians believe, Splogeh will smile on the handiwork of his faithful servants and say, "The promise of Sploge has been fulfilled."

The Moronian caste is supported by a large and obedient middle grouping known as the Marteans. The Marteans still work, but less and less do they produce anything; more of them labor in jobs which produce little but serve much. In any case, that is no longer their chief

PROBING POPULAR CULTURE

function for Splogian society. Now their primary mission is to consume the great glut of goods the Moronian owners have put at their
disposal. The Splogian economy depends upon the reliable surges of
spending which move the goods for the introduction of new goods, in
an endless cycle of Martean consumption. Everything is disposable,
and indeed it is the height of patriotism to dispose of things so that
new things can be bought and used and then thrown away. Sploge
could not spread without the splurge of shopping and spending on
stuff. A form of Moronian-sponsored propaganda called "advertising" whets and directs the Martean appetite for more stuff. They learn
and live by the motto, "I never saw a thing I didn't want." Their wants
are unlimited and insatiable, ever renewed by the appearance of new
things to want. To this end, the Marteans roam the roads in vehicles
foraging, in a strange version of the hunters and gatherers of old, for
the stuff that temporarily satisfies their craving. (It should be noted
that by every dietary measure, the Marteans are generally overweight
and out of shape, by literally stuffing themselves with food of the
most ghastly and unhealthy sort; but such gluttony is part of their heroic agenda to consume stuff.) The Marteans are an arterial people,
since their quest leads them to the places of purchase that line the superhighways. When they enter the markets and marts and malls, they
are possessed (a malady we term "glutomania") and move about as in
a trance glutifying their shopping carts and filling their vehicles with
stuff. However, the principle of disposability dictates that their new
toys will soon be used or broken or tired of, and will be discarded for
the next new thing.

To this end, the Marteans are committed to a program of waste.
The Localian dictum of "Waste not, want not" is replaced by the
Splogian motto, "Want not, so waste much." The Splogians long ago
realized that even though there is a finite amount that the Marteans
can consume, there is an almost infinite amount they can waste. So
waste becomes a measure of how well the consumption economy is
doing. The Moronian government has established a Department of
Waste Analysis, which measures just how much is being wasted. If an
increase occurs in the stuff thrown away, there is joy in ruling circles
because the Marteans are out searching for more stuff. Recognition is
given to those who are paragons of prolific wastefulness, and it is a
proud indicator of a community's prosperity as to how much stuff it
throws away. If the Moronian class specializes in conspicuous con-

sumption (consuming things that denote wealth, such as fashion, mansions, and fine wines), the Marteans are expected to engage in exhaustive consumption, which always challenges the limits of what is available and how much can be spent. For the corollary of expanding waste is increasing debt, the gross amount the Marteans owe in order to support the economy of spend-and-pitch-and-spend some more. To the extent the Marteans are deeply in debt, the Splogian economy is fundamentally sound. (For the anthropologist, this is a variation on the tribal ritual of "potlatch," whereby honors and accolades go to those who are the most lavish spendthrifts daring to take no thought of the morrow.) Debt is only one of the vices the Splogians encourage in order to further consumption. Indeed, Splogia is the foremost proponent of Mandeville's Principle (after the eighteenth-century English economist): to the extent a society insists on virtue, it is not prosperous; to the extent it promotes vice, it is prosperous. In Splogia, consumption keeps exploring new areas formerly deemed vice: gambling, pornography and the sex industry, spring break migrations, decadent spas for the rich. The vices of fun and play, often to wretched excess, create more prosperity through expanding the opportunities to Splurge!

The Moronians are at the top of Splogian society, and the Marteans at the vital center. But we note other distinct groupings at the margins, who are largely tolerated as either archaic residues or charming eccentrics. The Localians still exist on the margins, working in local jobs and living the local life. Many of them aspire to be Marteans, but lack the wealth to do it properly. So many of them dream of winning the lottery, and hope their children can someday join the Martean class, someday hopefully shopping till they drop. (The Localians often conceive heaven as a gigantic mart of endless cornucopia, the descendant of the "big rock candy mountain" of ancient lore.) Localia is not so much a place as it is a set of occupations, such as restaurant staffs and truckers and cashiers and small farmers. They are a disappearing society, who will take with them certain archaic skills, such as sewing, canning, and saving. Although the Moronian class may become helpless without their skills, the Localians are a doomed group, since they are politically impotent and socially invisible. Who will fix the plumbing and clean houses after their passage is unclear, but immigrants are a likely source of cheap labor for menial or mechanical tasks, freeing the consuming classes for their primary social mission of serving Splogeh by sanctified splurging.

Two marginal groups in Splogia threaten the *ethos* of unlimited splogation. The first group is called the Luddians, after their ancient tribal hero, a mysterious General Ned Ludd of yore. The Luddians are alienated from the prime directives of Splogian culture, even though they often exist within its structures. (Indeed, they are often criticized as hypocrites or phonies, damning the pollution of the environment while driving SUVs or condemning technology over the Internet.) But many Luddians are quite sincere, and make various efforts to simplify their lives and realize alternative values. Most are at the mercy of the Splogian system, but regard it as mindless and inane. Most Luddians must be content with small revolts, such as reading serious books, using manual typewriters, eating organic food, and sponsoring environmental causes and hiking trials. Their resistance may be futile, but nevertheless they conduct a kind of quixotic existence, waxing nostalgic for some aspects of Localian culture (e.g., baking bread, raising goats, gardening) and waning apocalyptic about the consequences of splogation. The Luddians are often well educated, and could conceivably become the critical mass for a political movement to overthrow the Moronian ruling class. But this would require a wider grouping, convincing the Localians that they are not helpless and the Marteans that they are not gluttons.

A far greater threat to Splogia comes from a nascent group we may term the Cyburbians. This new sector of society seems harmless enough, since they spend enormous amounts of time staring at computer screens, and whose sole recreation seems to be playing video games. Since they are not physically adept or socially aware, they might seem an unlikely source of political change. But they do seem to have a new god, Neteh, to worship, and generally seem to have a great deal of contempt for the intellectual density and informational incuriosity of the Moronians. Although it pleases Splogeh that they will shop on the Internet, they are not basically shoppers but rather seekers, looking for new and absorbing affiliations and analyses, some of them no doubt quite subversive. The worship of Neteh is a latter-day Gnostic religion, since their god exists in an ethereal world and invites the Cyburbians to live a discarnate life. Such a movement toward a new god and new life would seriously threaten the reign of Spolgia and the ascendancy of Splogeh. The Cyburbians can be seen in every office, and live on every street; they are everywhere, linked together through the electronic power of Neteh, and constitute a threat as a mystic faith and monastic order who may challenge the edicts of Splogeh. The Cyburbians, whose god lives nowhere and everywhere, might develop into a utopian movement that asks people to forsake the immediate and palpable things of Splogeh for the sacrosanct data of Neteh. Like the Christians of old, those who were slaves to consumption and accumulation might find a higher calling and purpose in the sacral realm of knowledge and erudition. Just as Splogia was based on ignorance, Cyburbia would be based on enlightenment, replacing impulse with inquiry. Eventually as an established church and perhaps state, Cyburbia could be the new sacred order of the twenty-first century, displacing Splogeh just as Zeus long ago became a mere curiosity of a discredited faith. The margin becomes the center, and the former center an anachronism, with people wondering how the Splogians ever could have lived that way. The possibility exists that the Splogians could suppress the worship of Neteh as a pernicious heresy, in which case the Cyburbians would be forced to communicate in secret, like the Christians of old meeting in catacombs. The Cyburbians would become adept at linking Neteh's followers through secret Web sites.

At the bottom and the fringes of Splogian society are the Vulgarians, rebellious or bohemian groups who are bothersome but not at the moment a social threat. Vulgarians include the homeless, beggars and hobos, gangs, and slum dwellers who are drugged into submission. But we may also note such vulgar groups of eternal party animals as surfers, bikers, barflies, and fraternity boys. Some Vulgarians become celebrated as enviable hell-raisers, as in movies about the biker groups or fraternity high jinks, and in the publicized antics of rock groups living the high life. Vulgarians constitute no coherent rebel group, but rather engage in hedonistic rebellions, or as with the homeless, occupy themselves with the desperate struggle for survival. Vulgarians may mock the Splogian ideal, but they exist on the extreme margins and thus are of little importance to the project of splogation.

That project is the fanatical mission of a group of zealots within the ruling Moronian class. They are deemed Imperians, whose chosen mission is no less than the splogation of the world. The Imperians are energized by the faith that assures them that they are serving the will of Splogeh, and believe that the world will not be perfected until the process of splogation has been expanded to its note. "Let us spread sploge everywhere!" they cry, and seek venues in every remote corner of the globe for unsplogated places. They seek to bring the gospel of Splogeh to Bedouin camel drivers, Amazonian Indians, Laotian peasants, Tibetan monks—indeed to all peoples and places that do not know the grace of Splogeh. They will not rest until malls adorn mountaintops, asphalt covers valleys, franchise food courts grace every hiking trail, and development has disrupted every community. Imperians see a day where superhighways have bulldozed through every town and neighborhood, McMansions line every lake and river and shore, and every forest and field have succumbed to sprawl. Only then will the world approach perfection, when everything natural has been replaced by the artificial and every resource has been used up. The empire of Splogia will then be complete. What happens after that is an incarnate mystery known only to Splogeh.

From the anthropological viewpoint, the Realm of Splogia is a new kind of society, whose glory and shame is in its commitment to rule of no limits. Splogia is virtually without precedent in human history, and it seems certain that future history will be radically altered by its success in expanding the principle of splogation everywhere. The Splogians overwhelm every archaic or fringe group, and are imbued with a messianic faith that spreads their way of life everywhere, enlisting everyone in the creed of Splurge! and every place in the beauty of Sprawl. But like all imperial people, they will meet mighty resistance in various quarters to their cultural dominion. The Moronian rulers have the assertive confidence of the unreflective, but their banishment of critical thought and conservative caution leads them into reckless expansions and interventions that court overreach and economic and political disaster. In the meantime, however, they are instructive to watch in action, for the consequences of their projects are felt wherever they have influence. Their expansive success depends upon the defeat of every value or practice that enjoins restraint, and contempt for critical inquiry that questions the set of platitudes invoked to justify Moronian rule and universal splogation. The ascendancy of Splogia teaches us that human wants may indeed be insatiable and infinitely malleable, regardless of the damage done to moral order, environmental sustainability, and even human health. The Splogians threaten to engulf the world, and the nature of that world when it is completely splogicized requires some imaginative forecasting. One very real possibility is splogicide, wherein the mania for splogation becomes so adrent and extensive that the Splogians become a "death culture," steadfastly committed to their cultural imperative to the extent that they destroy themselves by destroying everything else.

Anthropologists have long debated whether human nature is fixed or flexible, designed to be in harmony with nature or forever alienated from our natural condition. The Splogian conviction that nature is merely a resource to be used up and discarded suggests another idea: that people can become so committed to transcend nature as to create a completely artificial society without regard for natural or moral law. What has been abandoned in the process is the human base in nature and the tradition of natural law, and what is adopted is a power philosophy that sees the world in terms of conquest, defining everything and everyone as expendable. Splogia may conquer much of the world, but the awful and inevitable price may be the discovery that you tempt Mother Nature at your own risk.

My Students Speak

We are the generation of gadgets galore, instant everything, with no time to fix or eat a decent meal.

Lori Chisenhall, 2003

My students speak and I listen. They are not only from another generation (many are younger than my children), but also from another world. How can I probe popular culture if I don't listen and heed them?

Most of my students, like Lori Chisenhall, are studying popular culture and writing about it. She continues: "We revolve our lives around computers, the boob tube, Internet, brand names, commercialized this and mass-produced that. Is everybody happy? The answer is no."

Lori and her classmates watch channels, see films, read books that I never encounter. One of their current favorite films is *Fight Club,* in which leading character Tyler Durden speaks for the middle children of history, who (unlike my generation) have experienced no Great War or Great Depression. Their great war is spiritual; our great depression, our lives.

Reading through term papers and exams, I find a great deal of this kind of uneasiness and questioning in my students. Of course, it is not universal, and there is still some of the optimism and self-confidence that is the trademark of youth. Yet the ground on which we stand, and the doubts that plague us, will not go away.

We have something unlike other generations, they seem to agree: terrorism, right in our own backyards, perhaps even (with the sniper that terrified the Washington area in October 2002) at our gas pumps. "We are the children of terrorism," a student writes. "No matter how much we do to prevent it or fight it, fears will always be present. The monumental impact of 9/11 will forever be etched in our minds."

How to solve all this? "We cannot solve anything with a Coke," another student writes. "Wearing that pair of Abercrombie and Fitch jeans or driving a BMW M3 coupe won't make you who you are. Having the latest Palm or laptop, watching MTV and reality TV won't help you define yourself."

Sometimes the student tone is more one of resignation rather than indignation. "For now we shop. For now we talk on our cell phones, check our e-mails, and eat our microwave dinners. We are empty inside and need to fill the void with something. We go on because that is what we do and who we are."

Sometimes an ancient reference takes over. "We can no longer appreciate reality. Our reality has been replaced by Plato's cave. We have been in our cave since childhood, so we see no need for getting out and facing reality. In fact, many would question that there is an alternative to what we are shown by the media."

Realizing I am what some of my young students call "an ancient," I also see that they share my doubts about the rapid acceleration of technology. "Is it really giving us the Good Life, in what the technocrats like to call the Global Village? Or is it creating unbridgeable gaps between nations, religions, and economies—a digital divide? Well-to-do people are communicating and banking in the Global Village, but what about everyone else?" Another student comment takes on a more severe tone: "Will the rat race go on forever? Will this evolution simply be a process of renaming what has already been done? Our purpose remains unclear. All we can do is sit back and wonder: are we there yet?"

One student admits that we are consumed by labels and crowds. Still, she writes, we have evolved into a culture with more drive, competition, and a possibility of making dreams come true. She adds, "We have developed equal parent-sharing duties and a high self-worth."

Meredith Bauer takes another approach: nostalgia. This is how she begins her class notebook:

> Stroll with me. Close your eyes, and go back before the Internet, AIDS, herpes; before semiautomatics and crack; before SEGA or Nintendo. I'm talking about sitting on the curb, chocolate milk, going home for lunch, hopscotch, skates with keys, Cracker Jacks and hula hoops. When nearly everybody's mom was at home when the kids came home from school. When a quarter was a decent allowance. When you'd reach into a muddy gutter for a penny. *Mickey Mouse Club, Kukla, Fran and Ollie,* Dick

Clark's *American Bandstand* . . . all in black and white and mom made you turn the set off when a storm came. Paper chains at Christmas, silhouettes of Washington and Lincoln. What about the girl who dotted her i's with hearts?

In many different ways, and approaches, one senses the way fear haunts the children who grew up in the Clinton years. They are caught in cyberspace, trying always to catch up in "hurry-up time." "Yet you have to wonder," a bright communication studies major writes, "would being left behind be so terrible? Computers and machines are not what make us happy. It's the joy we get out of living in the amazing world around us. What scares me is that many people never even see it."

Others come to the defense of the omnipresent computer. "They give us more time, less hassle, and better performance. However, when the chores are done, we don't go outside to play hopscotch or flag football or hide-n-go-seek. Instead, we sit on the couch, cramming fattening food in our mouths, watching satellite TV, playing video games."

She goes on to tell a moving personal tale. She went home for the weekend, anxious to be with her family. She wanted one of Mom's good home-cooked meals, and a lively conversation about family matters. Called to dinner, she realized her mother was sitting in what had always been "her chair" for meals. "I didn't bother to ask any questions," she writes. "I assumed she wanted to sit closer to my father." Then she attempted, in vain, to start a conversation about her work at college. Both her mom and dad let out loud laughs and simply ignored her.

Suddenly she realized in a flash that "the family" had changed. Her parents were watching a television program behind her head. They had let television take over the dinner table and everything she came home for.

Epilogue

How Are We Different?

Thank God I am an American!

Daniel Webster

There was an American style before any European lived here. Today we are discovering exciting things, and giving much more attention to Native Americans. But it has had only a negligible effect on most Europeans who migrated to America. Where then does today's prevalent "style" begin? What is it like in our new century? Are we still style setters?

Style begins when humans occupy a land and make it their own. The first style on our land was that of people long called Indians, and now known as Native Americans. We do not know for sure where they came from or when, how they traveled, coped, and multipled. We do know that when the first Europeans arrived centuries later, they had highly integrated cultures that have existed up to the present. We know that they were driven from their native lands, slaughtered when they resisted superior technologies, and herded onto reservations. In recent years our understanding and respect for these tribal nations has increased greatly.

Hence our explanation of our national "style" has been seen as derived from Western Europe, especially nonconformist Britain. The colonies that eventually became the United States spoke English, read European books, and sold their raw materials to European markets. Our first president, General George Washington, wore a red British coat before donning the blue of the Continental Army. The noble words that Thomas Jefferson penned into our Declaration of Independence echoed those of English philosophers such as John Locke. The homes that Presidents Washington and Jefferson built for themselves—Mount Vernon and Monticello—copied those of European aristocrats, not native Americans.

It was one thing to win political independence and quite another to win cultural and social independence. As late as 1837 Emerson complained that we have listened too long to the "courtly muses of Europe; we kept listening. Then the situation reversed and Europeans were listening not to the courtly but the popular muses, the unembarrassed muses of jazz, mass production, Hollywood, shoot-'em-up Westerns, and Tin Pan Alley.

The famous formulation of this question came in the mid-nineteenth century. "What then is this new man?" de Tocqueville asked, "Who is this new man, this American?" We can still ask the question in the twenty-first century and raise a second question: Is America still, as Goethe believed, "the last best hope of the world?" Or will we be swallowed up in the Global Village? Has Ameristyle become technostyle? Are we best defined not by our old-style communities but our computers with their chat rooms? Is the divorce permanent?

When we thought of "settling America," we were traditionally taught about those brave and bold Europeans who followed Columbus, who "sailed the ocean blue in fourteen hundred and ninety-two" and "discovered" America. In fact, we paid little attention to the Native Americans already here, or to the Spaniards, who came earlier and colonized much of the New World.

Our accounts did, and still do, spotlight the relatively modest beginnings of the English at Roanoke Island, Jamestown, and Plymouth. Until the last generation, not only our formal "schoolbook" history, but also our popular culture, began there.

At the time of the seventeenth-century European invasion there were about a million Native Americans in North America (whom Columbus wrongly dubbed "Indians") and several million in Central and South America, where three advanced civilizations (Inca, Aztec, and Maya) flourished. They had sophisticated systems of irrigation, government, and mathematics—somewhat comparable to the high achievements in Sumer, Babylon, and Egypt. At the same time there were some unexpected absences: no wheels for transportation or machine purposes; no plows, oxen, or horses; no wheat, barley, or rice. Why was this the case? Why did these advanced civilizations disintegrate so quickly? Too little research has been done on these remarkable people, who seem to have been conquered not so much by invading armies as invading diseases, especially syphilis, measles, and smallpox.

So far, no evidence suggests that this vast continent was inhabited before 12,000 years ago, when settlers crossed over via the Bering Strait, then a land bridge. Other immigrants were moose, sheep, and bears, all having evolved in Asia. These "other Americans" were the first pioneers, who took the full force of vast and often hostile territory, survived, and flourished.

Their worst enemies were neither animals nor nature, but the later European invaders and their germs. Slaughtered, plundered, and pushed ever westward, the "Indians" would be given an unenviable tag line in American folklore: "The only good Indian is a dead Indian."

No doubt these Native Americans were often ferocious in resisting the invaders. They burned twenty-five towns between 1675 and 1677, during which time at least 800 colonists were killed. Allied with the French in the French and Indian War, they were deadly enemies. The seeds of hate and fear were sewn deeply. Years later, American popular culture was still seeking revenge. A 1922 Broadway play, *A Broadway Cowboy,* featured William Desmond who (according to the poster in the Odeon Theatre) "Killed six Indians after lunch each day." Ads assured patrons that the play was "peppy, spicy, and zippy."

Official culture had long been unsympathetic with the "savages." As early as 1838, President Andrew Jackson—in defiance of the Supreme Court—drove the Cherokees from their home in Florida westward on the "Trail of Tears" to preserve them from "civilization's infections"—and to confiscate their land. With different worldviews, the stronger Euro-American forces showed little sympathy for the natives. Treaty after treaty was made with the "Redskins" and then broken. In movie after movie these sons of Satan "bit the dust."

They were featured in stereotypic roles in books, magazines, and films. Few real natives either wrote or acted in these media. As late as 1984, in the 1982 made for television movie dealing with the native diaspora *(The Legend of Walks Far Woman),* the lead was given not to a Native American, but to Raquel Welch. Ugly battles arose over casino gambling rights given to Native Americans in the 1990s, and some of the old antagonisms lingered.

But the general attitude toward these "other Americans" has changed dramatically. Instead of being annihilated, the Indians' population is increasing, as are their supporters from the white population, thanks to Indian activists, the recent stress on New Age ecology, and the establishment of AIM (American Indian Movement) in 1968. AIM spearheaded the occupation of Alcatraz Island in 1969, the Bureau of Indian Affairs in 1969, and Wounded Knee in 1973. Hollywood got the message and began producing movies sympathetic to the long-despised Indians: *Tell Them Willie Boy Is Here* (1969), *Soldier Blue* (1970), *A Man Called Horse* (1970), and *Little Big Man* (1970). In the last of these, an Indian actor, Dan George, was actually given a major role as Old Lodge Skins. Nine years later Jay Silverheels, Tonto in the popular *Lone Ranger* television show, was the first native actor awarded a star on the Hollywood Walk of Fame. A Hollywood clone of *Little Big Man, Dances with Wolves* (1990) won seven Academy Awards, including Best Picture of 1990. *Geronimo: An American Legend* (1993) attempted to turn that popular outlaw into a genuine American folk hero. Walt Disney's *Pocahontas* (1995) attempted to revise and romanticize an early American myth. Pocahontas is turned into a Barbie doll figure—slim, smiling, loving, ethnic, and Native American—perfect for our time. Her mate (much younger in the film than real life) is Captain John Smith, "subjugator of nine and thirty kings," captured now with his head on the block for having killed two of Chief Powhatan's braves. Who can save him now?

Pocahontas, of course. So impressed was her father, Chief Powhatan, that he ordered Captain Smith released, to return to Jamestown and his heroic adventures. According to the myth, at least, she continued to visit Jamestown. Later on, by warning the settlers of an impending Indian attack, she saved the colony. Pocahontas became a Christian, was baptized, married John Rolfe, and sailed back to England as Lady Rebecca.

She got some of the adulation showered on a later twentieth-century princess, Diana. King James I presented her in court. London's Lord Bishop entertained Pocahontas with "festival, state and pompe." Suddenly stricken with smallpox, she died and is buried in the church in Gravesend.

We were long taught that America was "discovered" in 1492. But when was our popular culture "discovered"?

Not until the second quarter of the twentieth century, Reuel Denny concludes in a pioneering essay on that subject. This despite the fact

that much material we now accept as "popular" was produced and enjoyed for generations. Centuries would pass before popular culture
would be allowed to enter the sacred halls of academia. The earliest
reference to the subject didn't appear in the standard *Reader's Guide*
until 1960. To many elitists and critics, it is still "substandard."

Three and a half centuries earlier bushy-bearded Captain John
Smith had set the stage for popular memoirs. His *True Travels* created
an enduring legend of Pocahontas. Smith's *A True Relation* (1608)
was not only the first printed account of the Virginia settlement at
Jamestown, but the beginning of American popular literature. What
has not been discovered is "popular culture," but its early and contin-
ual impact on our national life, and that of much of the world.

In our century a new word keeps popping up to describe America:
transracial. We have seen the browning of mainstream America, and
drastic attacks on white supremacy, racism, or segregation, real or
perceived. The furor over Trent Lott, the Mississippi U.S. senator
who took these words too lightly in 2002, galvanized the high-profile
community.

There is wide acceptance of black superstars such as Eddie
Murphy, Michael Jordan, and Oprah, and mixed-race Tiger Woods.
Blacks were once derided, like Mohammad Ali, at one time the anti-
American draft-dodging ingrate, or George Foreman, once the brag-
ging bonebreaker, now the smiling hawker for car mufflers and home
grills. Get with it, man. Black women are going blonde or redhead,
and trendy white women are wearing dreadlocks. Whiteness is being
dissolved into a transracial mix. Everybody hears hip-hop.

Michael Lind predicted this about hip-hop in his 1995 book, *The
Next American Nation: The New Nationalism and the Fourth American Revolution.* Now advertisers are compelled to be multiracial. If it
isn't present in brand-name ads, we raise our eyebrows. The political
implications of this browning of America are enormous.

When an old mythology disintegrates, a new one originates—
along with new heroes. To discover a new mythos, we must create
and participate in it. That is what we must understand and promote.
Heroes, like Proteus of old, take on all manner of shapes and guises.
Our postmodern world has negated past heroes and created new environments of impersonal and invisible power—new patterns, new
space, new hype.

We find ourselves in a free-fall, borrowing, bending, and mending, having lost agreement not only about former heroes, but also about leaders, methods, and "the canon." The new technology has upended the old cosmos, moving us at the speed of light from reality to virtual or hyperreality. We live in Plato's cave, isolated by celluloid, cathode tubes, and clicking computers. The foundation of traditional authority has been deconstructed. All we are left with is a pile of rubble.

From that rubble we must construct another and better world. It may be more regional than global, more compassionate than competitive. The message is clear: change with the times, but never forsake our past. History must be revived and revivified. It must also be expanded to include those produced by our new emphasis on gender, race, and ethnicity.

Myths and legends that served us for centuries have disappeared in cyberspace. We must work hard to find both mission and meaning in our tumultuous time—look for and welcome new views of who we are, and who we want to be. This may not be as hard as some think.

The yearning for meaning—closely tied up in our myths, legends, and folklore—is always with us. Recall lines in the first book of the Bible, the Book of Genesis which told of giants and heroes.

The new heroes and heroines will be more than technocrats, sweepstakes winners, professional athletes, or get-rich-quickers. They will not depend, like so many of today's leaders, on spin doctors and doctored polls. They will tell the truth and spite the devil. They will stand for what made us "the land of the free and the home of the brave" in the first place.

Help wanted: heroes, heroines, and people of good please apply; people of depth, feeling, and vision. We have too many one-dimensional people who can see no further than their computer screen or the latest stock market figures. Is mass culture processing not only goods, but also people? Have we become commodities?

The future of our society rests in an understanding of our past: the eternal truths and strong communities. They have been degraded; but rebirth and renewal are able to recover what we seem to have lost. We cannot only renew, but create, a more human culture.

Sound transmitted by our new technology moves nearly a million times faster than it did a century ago; but it may not be a million times more important or useful. Recall what Henry David Thoreau said when asked to gather for the first telegraph message coming from

Texas to Massachusetts. When they have something important to tell us, he replied, he'd be anxious to listen. Can we say this as we confront unending spam, taped telephone solicitations, information overload? When can we demand it?

Hasten the time.

The central fact about American life is pluralism, the parent from which both localism and regionalism sprang. The region is an abbreviated symbol of national diversity. Regionalism, under whatever label, will survive long after outmoded nationalism and utopian globalism have disappeared.

Gone too will be arbitrary colonial boundaries Western nations imposed on Asia and Africa, ignoring regional and tribal boundaries developed over the centuries. What we say about American regionalism has global implications, for regionalism anywhere gives life its flavor and deep diversity. These differences are a major advantage, encouraging the interplay of viewpoints that foster flexibility and wisdom.

Allowing regions to cultivate their own genius and to find their happiness, sustenance, and security and pursuing these things is our best hope for a peaceful and productive twenty-first century.

Size and spectacle glitter and attract, but the enduring glory is in the minute—the small, specific, and immediate. Woodrow Wilson, Virginia born and bred, knew this. "A nation is as great, and only as great, as her rank and file," he wrote in his "New Freedom" domestic policies. "You can love a country if you begin by loving a community."

Walt Whitman, who became our greatest national poet by understanding the local and regional, put it best in *Song of Myself:*

> Who are you indeed who would talk or sing to America?
> Have you studied out the land, its idiom and people?
> Are you faithful to things? Do you teach what the land
> And sea, the bodies of men, womanhood . . . teach?
> Are you really of the whole People?

We can jet around the world and spend hours on the World Wide Web, but we eventually live on two soils. The first is physical, and from it come crops to sustain us. The second is spiritual and feeds another part of us. From it springs the tradition and mythology that cement us together.

To endure, a civilization must plant seeds in both soils. These must draw not only from the land, but from what love and deep feeling can make of those who cherish and work it. Here is the sinew of the soul, without which human meaning evaporates.

Two soils, two strengths. One strength is of the earth, the other of the human spirit. May they merge and sustain us in the tumultuous days ahead.

"The old order changeth, yielding to the New"—Alfred Lord Tennyson

Appendix

Essential Electronic Resources

News and Breaking Stories

CNN Interactive (links by story categories)
<http://www.cnn.com>

Conservative spin on the news at Town Hall
<http://www.townhall.com/>

Yahoo! News
<http://news.yahoo.com>

Historical and Cultural Organizations

The Popular Culture/American Culture Associations
<http://www.h-net.org/~pcaaca>
This Web site has book reviews, area chair names and addresses, program announcements, and lists of officers for the organizations. Membership form and meeting information online.

The American Studies Association
<http://www.georgetown.edu/crossroads/asainfo.html>
Contains the newsletter of ASA along with information about American Studies programs across the land. Past issues of *The American Quarterly* on line.

The Organization of American Historians
<http://www.oah.org>
Publishes the *Journal of American History* and has excellent research, job, query announcements updated on a regular basis.

The American Historical Association
<http://www.theaha.org>
Organization's detailed newsletter, calendar of events, national program. Purchase pamphlets and publications from the site.

Special Archives of Important Documents and Histories

The National Archives
<http://www.nara.gov/>

U.S. Department of State International Information Programs
<http://usinfo.state.gov>

The Library of Congress American Memory Site, available from the
LOC home page
<http://www.loc.gov>

Phil Landon's Site: Film, Literature, Popular Culture, History Links Galore

<http://userpages.umbc.edu/~landon/>
An abundance of sources in all relevant areas of American and cultural studies by a senior scholar.

Film Links

Film & History
<http://h-net.msu.edu/~filmhis>
This journal has thirty-three years of its tables of contents on the Web
site plus discussions of recent and significant films in relation to history.
Every university library should subscribe to this handy source, and
many around the world do just that.

Further Reading

With so broad a subject covering not only centuries but millennia, a reading list could become a large volume in its own right. *Books in Print,* for example, contains 1,659 entries under "Popular Culture," with twenty-two subdivisions. Of those no longer in print, there is no end. Hence we include here books to supplement key sources already listed in the Notes. Only books and articles of special interest have been selected, and the most recent titles have been given preference.

Anderson, Walter T. *Reality Isn't What It Used to Be.* New York: HarperCollins, 1990.

Ang, Ien. *Desperately Seeking the Audience.* London: Routledge, 1991.

Bailey, Brian J. *The Luddite Rebellion.* New York: New York University Press, 1998.

Baudrillard, Jean. *Forget Foucault.* New York: Semiotext(e), 1987.

———. *The Ecstasy of Communication.* New York: Semiotext(e), 1988.

———. *The Cool Provocateur.* New York: Verso Books, 1999.

Baughman, James L. *The Republic of Mass Culture.* Baltimore: Johns Hopkins University Press, 1992.

Bauman, Zygmunt. *Modernity and Ambivalence.* Cambridge: Polity Press, 1991.

Bennett, David, ed. *Multicultural states.* London: Routledge, 1998.

Bigsby, C.W.E., ed. *Superculture: American Popular Culture and Europe.* Bowling Green, OH: Bowling Green University Popular Press, 1975.

Blackham, H.J. *The Future of Our Past.* Amherst, NY: Prometheus Books, 1996.

Blau, Judith R. *The Shape of Culture: A Study of Contemporary Cultural Patterns in the United States.* New York: Cambridge University Press, 1989.

Bloom, Harold. *The American Religion: The Emergence of the Post-Christian Nation.* New York: Simon & Schuster, 1992.

Boas, George. *The History of Ideas: An Introduction.* New York: Charles Scribner's, 1969.

Boorstin, Daniel J. *The Image: A Guide to Pseudo-Events in America.* New York: Vintage Books, 1961.

Browne, Ray B. "Popular Culture As the New Humanities." *Journal of Popular Culture,* 17(4), Spring, 1984.

———. *Against Academia: The History of the Popular Culture Association/American Culture Association and the Popular Culture Movement, 1967-1988.* Bowling Green, OH: Bowling Green University Popular Press, 1989.

Browne, Ray B. and Browne, Pat. *Digging into Popular Culture.* Bowling Green, OH: Bowling Green University Popular Press, 1991.

Browne, Ray B. and Fishwick, Marshall W., eds. *The Global Village: Dead or Alive?* Bowling Green, OH: Bowling Green University Popular Press, 1999.

Browne, Ray B., Fishwick, Marshall W., and Browne, Kevin O., eds. *Dominant Symbols in Popular Culture.* Bowling Green, OH: Bowling Green University Popular Press, 1990.

Burke, Peter. *Popular Culture in Early Modern Europe.* New York: Harper Torchbooks, 1978. See especially Appendix A: "The Discovery of the People: Select Studies and Anthologies, 1760-1846."

Callaghan, Colleen. "Looking Forward to the 21st Century." In *An American Mosaic,* Marshall Fishwick, ed. New York: American Heritage Custom Publishing, 1996.

Cantor, George. *Pop Culture Landmarks.* Gale Research, Inc., 1994.

Cantor, Norman F. and Werthman, Michael S., eds. *The History of Popular Culture.* New York: Macmillan, 1968.

Carlson, Lewis, ed. *American Popular Culture at Home and Abroad.* Lansing, MI: New Issues Press, 1995.

Carney, George O., ed. *Fast Food, Stock Cars, and Rock-n-Roll: Place and Space in American Pop Culture.* London: Rowman & Littlefield, 1995.

Chambers, Iain. *Popular Culture.* London: Routledge, 1986.

Chaucer, Geoffrey. *The Canterbury Tales.* New York: Oxford University Press, 1998.

Chowdhury, Omar. *Prelude to Bangladesh.* New Delhi: South Asian Publishers, 1980.

Chrisp, Peter. *The Parthenon.* Austin, TX: Raintree/Streck Vaughan, 1997.

Collins, Jim. *Uncommon Cultures: Popular Culture and Post-Modernism.* London: Routledge, 1989.

Combs, James. "The Play World of the New Millennium." In *Preview 2001+* (pp. 69-82), Ray B. Browne and Marshall W. Fishwick, eds. Bowling Green, OH: Bowling Green University Popular Press, 1995.

Combs, James and Nimmo, Dan. *The Comedy of Democracy.* Westport, CT: Praeger, 1997.

Cook, B.F. *The Elgin Marbles.* London: British Museum Press, 1997.

Crowley, David and Mitchell, David, eds. *Communication Theory Today.* Stanford, CA: Stanford University Press, 1994.

Daly, Steven and Wice, Nathaniel. *Alt. Culture: An A-to-Z Guide to the 90's.* New York: Harper Perennial, 1995.

Davis, Robert C. *The War of the Fists: Popular Culture and Public Violence in Late Renaissance Venice.* New York: Oxford University Press, 1994.

December, John, ed. *The World Wide Web Unleashed.* Indianapolis, IN: SAMS, 1995.

Douglas, Ann. *The Feminization of American Culture*. New York: Farrar, Straus, & Giroux, 1998.

Duffy, Dennis. *Marshall McLuhan*. Toronto: McClelland and Stewart, 1969.

During, Simon, ed. *The Cultural Studies Reader*. New York: Routledge, 1993.

Eco, Umberto. *A Theory of Semiotics*. Bloomington: Indiana University Press, 1976.

Ferguson, Marjorie and Golding, Peter, eds. *Cultural Studies in Question*. London: Sage Publications, 1997.

Ferguson, Niall. *Empire: The Rise and Demise of the British World Order and the Lessons for Global Power*. New York: Basic Books, 2003.

Firth, Raymond. *Symbols: Public and Private*. Ithaca, NY: Cornell University Press, 1973.

Fishwick, Marshall W. *Common Culture and the Great Tradition*. Westport, CT: Greenwood Press, 1982.

———. *Seven Pillars of Popular Culture*. Westport, CT: Greenwood Press, 1985.

———. *Go, and Catch a Falling Star: Pursuing Popular Culture*. New York: American Heritage, 1994.

———. *An American Mosaic: Rethinking American Culture Studies*. New York: American Heritage, 1996.

Fiske, John. *Understanding Popular Culture*. Boston: Unwin Hyman, 1989.

Fiske, John. *Media Matters: Everyday Culture and Political Change*. Minneapolis: University of Minnesota Press, 1994.

Frow, John. *Cultural Studies and Cultural Value*. Oxford: Clarendon Press, 1995.

Gans, Herbert. *Popular Culture and High Culture*. New York: Basic Books, 1974.

García Canclini, Nestor. *Transforming Modernity: Popular Culture in Mexico*. Austin: University of Texas Press, 1995.

Gates, Bill. *The Road Ahead*. New York: Viking Penguin, 1996.

———. *The Secret Diary of Bill Gates*. New York: Andrews McMeel, 1998.

Giroux, Henry A. *Disturbing Pleasures: Learning Popular Culture*. New York: Routledge, 1994.

Gitlin, Todd, ed. *Watching Television*. A Pantheon Guide to Popular Culture. New York: Pantheon Books, 1986.

Glassie, Henry. "Skill." In *The Old Traditional Way of Life,* Robert E. Walls and George Schoemaker, eds. Bloomington: Trickster Press, 1989.

———. *The Spirit of Folk Art*. New York: Harry N. Abrams, 1995.

———. *Turkish Traditional Art Today*. Bloomington: Indiana University Press, 1996.

———. *Art and Life in Bangladesh*. Bloomington: Indiana University Press, 1998.

Goldstein, Jeffrey, ed. *Why We Watch: The Attractions of Violent Entertainment*. New York: Oxford University Press, 1998.

Goodwin, Andrew. *Dancing in the Distraction Factory: Music, Television, and Popular Culture*. Minneapolis: University of Minnesota Press, 1993.

Gordon, Ian. *Comic Strips and Consumer Culture, 1890-1945.* Washington, DC: Smithsonian Institution Press, 1998.

Gray, George Z. *The Children's Crusade.* Stark, KY: DeYoung Press, 1997.

Griswold, Wendy. *Cultures and Societies in a Changing World.* Thousand Oaks, CA: Pine Forge Press, 1994.

Grossberg, Lawrence. *Bringing It All Back Home: Essays on Cultural Studies.* Durham, NC: Duke University Press, 1997.

Grunig, James E. "Turning McLuhan on His Head." In *The Global Village: Dead or Alive?* (pp. 14-23), Ray B. Browne and Marshall Fishwick, eds. Bowling Green, OH: Bowling Green University Popular Press, 1999.

Hall, Peter Dobkin. *The Organization of American Culture, 1700-1900.* New York: New York University Press, 1984.

Hooper, J.T. and Burland, Cottie A. *Art of Primitive People.* Philadelphia: Philosophical Library, 1954.

Hunter, James D. *Culture Wars: The Struggle to Define America.* New York: Basic Books, 1991.

Inglis, Fred. *Cultural Studies.* Oxford: Blackwell Publishers, 1993.

Jacoby, Russell. *The Last Intellectuals: American Culture in the Age of Academe.* New York: Basic Books, 1987.

Jayaraman, Raja. "Border and Borderless Culture: A Study of the Process of Recreation and Maintenance of Ethnic Boundary in a Global Society." In *The Global Village: Dead or Alive?* (pp. 145-163), Ray B. Browne and Marshall W. Fishwick, eds. Bowling Green, OH: Bowling Green University Popular Press, 1999.

Jensen, Richard. "Internet 2001 and the Future." In *Preview 2001+* (pp. 209-216), Ray B. Browne and Marshall W. Fishwick, eds. Bowling Green, OH: Bowling Green University Popular Press, 1995.

Johnson, B.L.C. *Bangladesh.* New York: Barnes and Noble, 1975.

Jones, Joel. "American Studies: The Myth of Methodology." In *The American Self* (pp. 26-39), Sam Girgus, ed. Albuquerque, NM: University of New Mexico Press, 1981.

Joyce, Patrick. *Visions of the People.* Cambridge: Cambridge University Press, 1994.

Kaes, Anton. *From Hitler to Heimat: The Return of History As Film.* Cambridge, MA: Harvard University Press, 1989.

Kaufmann, Roland G. *King Tut's Reality: The Seven Steps to Redemption.* Las Vegas, NV: Heridonius Foundation, 1996.

Kittelson, Mary L. *The Soul of Popular Culture.* Chicago: Open Court Publishing, 1997.

Kroeber, A.L. and Kluckhohn, Clyde. *Culture: A Critical Review of Concepts and Definitions.* New York: Harper & Row, 1963.

Landau, Elaine. *The Curse of Tutankhamen.* Brookfield, CT: Millbrook Press, 1996.

Landay, Lori. *Madcaps, Screwballs, and Con Women.* Philadelphia: University of Pennsylvania Press, 1998.

Lapham, Lewis H. *Waiting for the Barbarians.* New York: Verso Books, 1997.

Lash, Scott and Friedman, James, eds. *Modernity and Identity.* Oxford: Blackwell Publishers, 1992.

Levinson, David and Christensen, Karen. *The Global Village Companion.* New York; ABC-Clio, 1996.

Maloni, Kelly, Wice, Nathaniel, and Greenman, Ben. *Net Chat.* New York: Random House, 1994.

Marchand, Philip. *Marshall McLuhan: The Medium and the Messenger.* Cambridge, MA: MIT Press, 1998. [Between 1951 and 1995, twenty of Marshall McLuhan's (1911-1982) books were published, as well as seventeen books about him, plus scores of articles and interviews. They are all listed in Marchand's biography. Both friend and mentor, McLuhan had more influence on my work than any other person. The most influential three books were *The Gutenberg Galaxy: The Making of Typographic Man* (1962); *Understanding Media: The Extensions of Man* (1964); and *War and Peace in the Global Village* (1968).]

McCurdy, Howard E. *Space and the American Imagination.* Washington: Smithsonian Institution Press, 1997.

McRobbie, Angela. *Postmodernism and Popular Culture.* New York: Routledge, 1994.

Meyrowitz, Joshua. *No Sense of Place: The Impact of Electronic Media on Social Behavior.* New York: Oxford University Press, 1985.

Mitroff, Ian I. and Bennis, Warren. *The Unreality Industry.* New York: Oxford University Press, 1989.

Monaco, James. *Celebrity: The Media As Image Makers.* New York: Dell, 1978.

Moore, R. Laurence. *Selling God: American Religion in the Marketplace of Culture.* New York: Oxford University Press, 1994.

Morgan, David. *Visual Piety.* Berkeley: University of California Press, 1998.

Morgan, Hal. *Symbols of America.* New York: Penguin Books, 1987.

Murfin, Ross and Ray, Supryia M. *The Bedford Glossary of Critical and Literary Terms.* Boston: Bedford Books, 1997.

Neal, Arthur G. "Cultural Fragmentation in the 21st Century." In *Preview 2001+* (pp. 111-126), Ray B. Browne and Marshall W. Fishwick, eds. Bowling Green, OH: Bowling Green University Popular Press, 1995.

Neuberg, Victor E. *Popular Literature: A History and Guide.* Middlesex, UK: Penguin Books, 1977.

Nevitt, Barrington and McLuhan, Maurice, eds. *Who Was Marshall McLuhan?* Toronto: Stoddart, 1995.

Nolan, Riall W. *Building the Global Village.* New York: NAL Dutton, 1999.

Novak, Robert. *King Tut in America.* Fort Wayne, IN: Windless Orchard, 1988.

Nye, Russel. *The Unembarrassed Muse: The Popular Arts in America.* New York: Dial Press, 1970.

Oriard, Michael. *Reading Football: How the Popular Press Created an American Spectacle.* Chapel Hill: University of North Carolina Press, 1998.

Palagia, Olga. *The Pediments of the Parthenon.* New York: E. J. Brill, 1993.

Phy, Allene Stuart, ed. *The Bible and Popular Culture in America.* Philadelphia: Fortress Press, 1984.

Polhemus, Ted and Procter, Lynn. *Pop Styles.* London: Vermilion, 1984.

Postman, Neil. *Amusing Ourselves to Death: Public Discourse in the Age of Show Business.* New York: Viking, 1985.

———. *Technopoly: The Surrender of Culture to Technology.* New York: Vintage Books, 1993.

Prose, Francine. *Primitive People.* New York: Ivy Books, 1993.

Real, Michael. *Super Media.* London: Sage, 1989.

———. *Exploring Media Culture.* Thousand Oaks, CA: Sage, 1996.

Rifkin, Jeremy. *The End of Work.* New York: Putnam, 1995.

Robinson, Frank M. and Davidson, Lawrence. *Pulp Culture: The Art of Fiction Magazines.* Portland, OR: Collectors Press, 1998.

Rollins, Peter C., ed. *Hollywood As Historian.* Lexington: University Press of Kentucky, 1998.

Rollins, Peter C. and O'Connor, John E., eds. *Hollywood's World War One.* Bowling Green, OH: Bowling Green University Popular Press, 1997.

———. *Hollywood's Indian.* Lexington: University Press of Kentucky, 1999.

Rosenberg, Harold. "Philosophy in a Pop Key." *The New Yorker,* February 27, 1965.

Rosenstone, Robert A. *Visions of the Past: The Challenge of Film to Our Idea of History.* Cambridge, MA: Harvard University Press, 1995.

Roszak, Theodore. *The Cult of Information: The Folklore of Computers and the True Art of Thinking.* New York: Pantheon, 1986.

Rowe, David. *Popular Cultures: Rock Music, Sports, and the Politics of Pleasure.* London: Routledge, 1995.

Rubenstein, Ruth P. *Dress Codes: Meanings and Messages in American Culture.* Boulder, CO: Westview Press, 1995.

Rushkoff, Douglas. *Cyberia: Life in the Trenches of Hyperspace.* New York: Ballantine, 1992.

———. *The GenX Reader.* New York: Ballantine, 1994.

———. *Media Virus: Hidden Agendas in Popular Culture.* New York: Ballantine, 1994.

Sacks, Peter. *Generation X Goes to College.* Chicago: Open Court Press, 1996.

Sanderson, George and MacDonald, Frank. *Marshall McLuhan: The Man and His Message.* Golden, CO: Fulcrum, 1989.

Sandos, James A. and Burgess, Larry E. *The Hunt for Willie Boy: Indian-Hating and Popular Culture.* Norman: University of Oklahoma Press, 1994.

Savage, William W. Jr. *Commies, Cowboys, and Jungle Queens.* Amherst, MA: University Press of New England, 1988.

Schechter, Harold and Semeiks, J.G. *Patterns in Popular Culture*. New York: HarperCollins, 1980.

Schwartz, Richard A. *Cold War Culture: Media and the Arts, 1947-1990*. New York: Facts on File, 1997.

Scribners Reference Shelf. *Encyclopedia of American Cultural History*. New York: Scribners, 1999.

Shaw, Bud. "The Mission Football," *Cleveland Plain Dealer,* October 10, 1993, 1D.

Shuman, Michael. *Towards a Global Village*. London: Pluto Press, 1994.

Siddiqui, Ashraf. *Folkloric Bangladesh*. Dhaka: Eden Press, 1977.

Smoodin, Eric, ed. *Disney Discourse: Producing the Magic Kingdom*. New York: Routledge, 1994.

Soccolich, R.M. *100 Steps Necessary for Survival in the Global Village*. New York: Seaburn Books, 1997.

Springhall, John. *Youth, Popular Culture, and Moral Panics: Penny Gaffs to Gangsta-Rap, 1830-1996*. New York: St. Martin's Press, 1998.

Storey, John. *An Introductory Guide to Cultural Theory and Popular Culture*. Athens: University of Georgia Press, 1993.

Swiatecka, M. Jadwiga. *The Idea of the Symbol*. Cambridge: Cambridge University Press, 1980.

Taylor, Timothy. *Global Pop*. London: Routledge, 1996.

Tindall, W.Y. *John Bunyan, Mechanick Preacher*. New York: Macmillan, 1934.

Toynbee, Arnold. *Civilization on Trial*. New York: Oxford University Press, 1948.

U.S. Department of State. *Post Report on Bangladesh*. Washington, DC: Government Printing Office, 1981.

van Elteren, Mel. "The Complexities of Cultural Globalization Revisited." In *The Global Village: Dead or Alive?* (pp. 36-56), Ray B. Browne and Marshall W. Fishwick, eds. Bowling Green, OH: Bowling Green University Popular Press, 1999.

Varnedoe, Kirk and Gopnik, Adam. *High and Low: Modern Art, Popular Culture*. New York: Museum of Modern Art, 1990.

Webster, Duncan. *Looka Yonder!: The Imaginary America of Populist Culture*. London: Routledge, 1988.

Wellman, Barry, ed. *Networks in the Global Village*. Boulder, CO: Westview Press, 1995.

Williams, Raymond. *The Sociology of Culture*. Chicago: University of Chicago Press, 1995.

———. *The Politics of Modernism*. London: Verso Books, 1996.

———. *Problems in Materialism and Culture*. New York: Verso Books, 1997.

Willis, Susan. *A Primer for Daily Life*. London: Routledge, 1992.

Yoder, Don. *Discovering American Folklife*. Mechanicsburg, PA: Stackpole Books, 2001.

Zeldin, Theodore. *An Intimate History of Humanity*. New York: HarperCollins, 1995.

Zinsser, William. *Pop Goes America*. New York: Harper & Row, 1966.

Notes

The Probing Process

1. Peter Burke, *Popular Culture in Early Modern Europe* (New York: Harper & Row, 1978, and Robert Redfield, *Peasant Society and Culture* [Chicago: University of Chicago Press, 1956]).

2. *Funk and Wagnall's Standard Dictionary of Folklore, Mythology and Legend,* edited by Marcus Leach, 2 volumes (New York: Funk and Wagnall, 1949-1950).

3. See Richard Gid Powers and Hidetoshi Kato (Eds.), *Handbook of Japanese Popular Culture* (Westport, CT: Greenwood Press, 1989).

4. In our age of the monograph, few authors try for the scope and depth which I suggest we need now. My favorite try is Max Lerner's *America As a Civilization* (New York: Simon and Schuster, 1957). My own much more modest attempt was published as *Common Culture and the Great Tradition* (Westport, CT: Greenwood Press, 1982).

The American Studies Link

1. Irwin Unger, "The 'New Left' and American History: Some Recent Trends in United States Historiography," *American Historical Review,* LXXII, July 1967, p. 124.

2. Sigmund Skard, *American Studies in Europe* (Philadelphia: University of Pennsylvania Press, 1954), p. 122.

3. The Harvard program dated from 1936, but the real expansion came after 1946. See Robert Walker, *American Studies in the United States* (Baton Rouge: Louisiana State University Press, 1958).

4. See Board of Foreign Scholarships, *International Educational Exchange: The Opening Decades, 1946-1966* (Washington, DC: U.S. Government Printing Office, 1966, 0-227-952); and Sigmund Skard, *American Studies in Europe.*

5. *New York Review* for July 15, 1965.

6. Henry Nash Smith, "Can 'American Studies' Develop a Method?" The essay is reprinted in Joseph Kwiat and Mary Turpie, *Studies in American Culture* (Minneapolis: University of Minnesota Press, 1960, p. 94).

7. See John Higham, "The Cult of the American Consensus," *Commentary,* XXVII, February 1959, pp. 93-100; William A. Williams, *The Contours of American History* (New York: World Publishing Company, 1961); and the journal *Studies on the Left.*

8. Howard Mumford Jones, *Education and World Tragedy* (Cambridge, MA: Harvard University Press, 1947).

9. Susan Sontag, *Against Interpretation* (New York: Farrar, Straus, and Giraux, 1966), p. 298.

10. M. W. Fishwick interview with John M. Culkin in *Pace* for September 1967, pp. 12-13.

11. For an example of such a plea, see Lawrence Chisolm's essay on "Cosmotopian Possibilities," in Marshall Fishwick (Ed.), *American Studies in Transition* (Philadelphia: University of Pennsylvania Press, 1964).

The East-West Pop Link

1. *The Pacific Rim and the Western World* (Boulder, CO: Westview Press, 1987), edited by Philip West and Frans A. M. Atling von Geusau, provides an up-to-date series of strategic, economic, and cultural perspectives. *Pacific Basin and Oceania,* edited by Gerald Fry and Rufino Mauricio in the "World Bibliographical Series" (Santa Barbara, CA: Clio Press, 1987) is the most comprehensive bibliography to date, annotating 1,178 books covering a multitude of subjects and places. *The Emerging Pacific Community: A Regional Perspective,* edited by Robert L. Downen and Bruce J. Dickson (Boulder, CO: Westview Press, 1984) examines the advantages and disadvantages of this concept.

2. For more details, see the 1987 *Asia-Pacific Report,* edited by Charles E. Morrison and issued by the East-West Center (Honolulu: University of Hawaii Press, 1988).

3. For a fuller development of this point, see Alan Gowans's *The Unchanging Arts* (New York: Oxford Press, 1974); and Peter Burke, *Popular Culture in Early Modern Europe* (New York: Harper and Row, 1978).

The Great Tradition

1. This is a vast and complicated subject, and sources are endless. I have been most helped by two books by Peter Laslett: *The World We Have Lost* (New York: Scribners, 1965) and *Household and Family in Past Times* (New York: Cambridge University Press, 1972). See also Gregory Baum, ed., *Sociology and Human Destiny* (New York: Seabury Press, 1980) and T. S. Eliot, *Tradition and the Individual Talent* (New York: Harcourt Brace, 1936).

Three excellent accounts of the Great Tradition are Morris Bishop, *The Penguin Book of the Middle Ages* (New York: Penguin, 1990); Werner Rosener, *Peasants in the Middle Ages* (Urbana: University of Illinois Press, 1992); and Georges Duby, *Love and Marriage in the Middle Ages* (Chicago: University of Chicago Press, 1994). For more details and insight, see Arno Borst, *Lebensformen im Mittelalter* (Frankfurt/Berlin: Ullstein, 1973); Hans-Werner Goetz, *Life in the Middle Ages: From the Seventh to the Thirteenth Century* (Notre Dame: University of Notre Dame Press, 1993); Lynn T. Courtenay, ed., *Engineering of Medieval Cathedrals* (Brookfield, VT: Ashgate, 1997).

Spoiled by Success?

1. Marshall Fishwick, *The World of Ronald McDonald* (Bowling Green, OH: Bowling Green University Popular Press, 1978). See also George Ritzer, *The McDonaldization of Society* (Thousand Oaks, CA: Pine Forge Press, 1993).

2. The first major study, Max Boas and Steve Chain's *Big Mac: The Unauthorized Story of McDonald's* (New York: Dutton, 1976), is not only unauthorized but bombastic and inaccurate. The independent journalist John F. Love spent four and a half years researching and writing *McDonald's: Behind the Arches* (New York: Bantam Books, 1986). It is much more reliable, if largely laudatory. The annual reports issued by the McDonald's corporation are essential for any study. For a wider scope and coverage, see Waverly Root and Richard de Rochemont, *Eating in America: A History* (New York: William Morrow, 1976).

3. Ray Kroc (with Robert Anderson) told his own story in *Grinding It Out: The Making of McDonald's* (Chicago: Henry Regnery, 1977).

4. John F. Love, *McDonald's: Behind the Arches,* pp. 1-9.

5. Alan Hess, "The Origins of McDonald's Golden Arches," *Journal of the Society of Architectural Historians,* XLV (March), 1986, pp. 60-62.

6. "Ray and Ronald girdle the globe," Marshall Fishwick, *Journal of American Culture,* 18(Spring 1995): 13-29.

7. See note 1. Whether McDonald's can adapt, service, and expand in the twenty-first century is an interesting question. See Greg Crister, *Fat Land: How Americans Became the Fattest People in the World* (New York: Houghton Mifflin, 2003), which exposes a calorie count of all McDonald's items and ends up suggesting a crusade: "Something Must be Done!" Health advocates are mounting a campaign reminiscent of the tobacco wars.

The Cowboy and World Mythology

1. Will James, *The Drifting Cowboy* (New York: Holt, 1943), p. 67.

2. Ray A. Billington, *Westward Expansion: A History of the American Frontier* (New York: Macmillan, 1949), p. 684.

3. For more details, read *The Future of the Great Plains: Report of the Plains Commission* (Washington, DC: Government Printing Office, 1936).

4. Herb Fagen, *Duke, We're Glad We Knew You* (Secaucus, NJ: Carol Publishers, 1998); Lee Pfeiffer, *True Grit* (Secaucus, NJ: Carol Publishers, 1998); Ronald Davis, *Duke: The Life and Image of John Wayne* (Norman: University of Oklahoma Press, 1998); Kinky Friedman, *God Bless John Wayne* (New York: Bantam, 1996); Garry Wills, *John Wayne's America: The Politics of Celebrity* (New York: Simon and Schuster, 1998); and Emanuel Levy, *John Wayne: Prophet of the American Way of Life* (Lanham, NY: Scarecrow Press, 1998).

5. Jon Scieszka, *The Good, the Bad, and the Goofy* (New York: Viking Press, 1992); Guy Logsdon (Ed.), *The Whorehouse Bells Were Ringing: And Other Songs Cowboys Sing* (Champagne, IL: University of Illinois Press, 1989); John White's *Git Along, Little Dogies* (Champagne, IL: University of Illinois Press, 1989); Katie Lee's *Ten Thousand Goddamn Cattle* (Jerome, AZ: Katydid Publishers, 1985); and Rick Steber's series, *Tales of the Wild West* (New York: Bonanza Press, 1988).

6. Sherwood Anderson, *A Story Teller's Story* (New York: B. W. Huebsch, 1935, p. 88).

7. Robert Warshow, "Movie Chronicle: The Westerner," *Partisan Review,* April 1954, p. 196.

Paul Bunyan: Fakelore Meets Folklore

1. As my chapter indicates, Paul Bunyan belongs to an earlier generation of folk-fake heroes, reaching his zenith before and in the 1940s. He is archetypal, the fore-runner of the many electronic superheroes and monsters that dominate the scene now. Still, Paul is fondly remembered. Daniel Hoffman's *Paul Bunyan: Last of the Frontier Demigods* has been republished several times, most recently in 1999 by the Michigan State University Press in East Lansing. D. Laurence Rogers published *Paul Bunyan: How a Terrible Timber Feller Became a Legend* in 1993; Marianne Johnston published her *Paul Bunyan* (American Legends) in 2001 (New York: Powerside Press); and Sandra Becker published *Paul Bunyan* (Folk Heroes) (New York: Weigl Education Publishers) in 2003. He remains part of our popular culture.

The Sign of the T: Henry Ford

1. The name "Ford" resounds not only through the land, but around the world. We all "watch the Fords go by." Henry Ford left books, museums, and factories as his legacy; dozens of writers, philosophers, and filmmakers have focused on him. I have drawn on them. The most helpful books were Keith Sward's *The Legend of Henry Ford* (1948) and Allan Nevins's *Ford: The Times, the Man, the Company* (1954); and the classic film, Charles Chaplin's *Modern Times* (1936). I got many nicknames for Model Ts from old-timers who owned one as well as from Ford deal-ers.

2. Among recent Ford books, these three were useful: Ford Bryan, *Beyond the Model T: The Other Ventures of Henry Ford* (Detroit: Wayne State University Press, 1990); Ray Batchelor, *Henry Ford: Mass Production, Modernism, and Design* (New York: Manchester University Press, 1994); and Neil Baldwin, *Henry Ford and the Jews: The Mass Production of Hate* (New York: Public Affairs, 2001). Fords are rolling off the assembly line as the new century begins.

3. John Dos Passos, *The Big Money* (New York: Harcourt Brace, 1926), p. 178.

Folk-Joke: Joe Magarac

1. Folk-jokes are shaped by both the culture and media; both change rapidly. By training and inclination, I favor print media as my source and evidence. In our new century, I would also have to look for pop-jokes on television, Internet, and e-mail. The most diligent and thorough of Magarac scholars, Hyman Richman, published "The Saga of Joe Magarac" in the *New York Folklore Quarterly* in Winter 1953. There is yet to be an equivalent electronic media work. A much wider audience heard of Joe in Jules B. Billard's "Ever Hear of Joe Magarac?" in the *Saturday Evening Post,* CCIXX, for February 22, 1947, p. 41f.

2. This story appeared in various Pennsylvania newspapers in October 1953.

3. See *U.S. Steel News,* "This Side of the Iron Curtain and Glad of It," XIII, No. 3, (July 1948), 2-5; and the comic book "Joe, the Genie of Steel" (Carnegie-Illinois Steel Corporation, 1950).

4. See Owen Francis, "The Saga of Joe Magarac: Steelman," in *Scribners Magazine,* XC (November, 1931), 505-511. The next month Francis published in the same magazine the story of Joe Zimmich, obviously another try at the same target. In it Zimmich performs great feats, works as hard as twenty men in the steel business, and helps his mill beat the others in production. Finally, a slag box falls on him and kills him, perhaps because the Zimmich story had no love angle—and failed to mention the magic word "folklore."

5. See F. J. Kern's *English-Slovene Dictionary* (Cleveland, 1919), P. A. Hrobak's *English-Slovak Dictionary* (Middletown, Pennsylvania, 1944), and Jindrich Prochazka's *English-Czech and Czech-English Dictionary* (Orbis Prague, 1950).

6. Mrs. Owen Francis maintains that her husband (now deceased) was in on the joke all along; but in light of what he wrote, this seems improbable.

ARF

1. See Witold Rybczynski, *How the Other Half Builds,* Paper No. 9, December 1984 from the Centre for Minimum Cost Housing, McGill University, Montreal, Canada. Other interesting points are set forth by Dennis Alan Mann, "Where Architecture and Popular Culture Diverge," in *Symbiosis: Popular Culture and Other Fields,* edited by Ray B. Browne and Marshall W. Fishwick (Bowling Green, OH: Bowling Green Popular Press, 1988), p. 177f.

2. James Marston Fitch, *Architecture and the Esthetics of Plenty* (New York: Columbia University Press, 1961), p. 20.

3. Lewis Mumford, "Megalopolis As Anti-City," *Architectural Record,* December 1962, pp. 101-108.

4. Robert Venturi and Denise Scott Brown, "Ugly Is Beautiful: The Main Street School of Architecture," *Atlantic Monthly,* April 1973, p. 37.

5. Alan Gowans, "Popular Arts and Historic Artifact," in Marshall W. Fishwick and Ray B. Brown (Eds.), *Popular Architecture* (Bowling Green, OH: Bowling Green Popular Press, 1975), p. 104.

6. See Donald D. Egbert, *Social Radicalism and the Arts, Western Europe: A Cultural History from the French Revolution to 1968* (New York: Knopf, 1970).

7. See Conrad P. Kottak, "Rituals at McDonalds," in *Ronald Revisited: The World of Ronald McDonald,* Marshall W. Fishwick (Ed.) (Bowling Green, OH: University Popular Press, 1983).

8. The question was raised as early as 1933 by Nicolai Hartmann in *Das Problem des Geistigen Seins: Intersuchungen zur Grundlegung der Geschichtsphilosophie und der Geistewissenschaften* (Berlin: W. deGruyter & Co.). More recent probes include Robert Dubin, *Theory Building* (New York: Basic Books, 1969) and Claude Levi-Strauss, *The Savage Mind* (Chicago: University of Chicago, 1970). We still know little of model-building as it applies to popular culture.

9. Sibyl Moholy-Nagy, *Native Genius in Anonymous Architecture* (New York: Horizon, 1957), p. 43.

10. Ada Louise Huxtable, "Only the Phony Is Real," *The New York Times,* May 13, 1973, p. 88.

11. Quoted by Russel B. Nye, *The Unembarrassed Muse* (New York: Dial, 1970), p. 378.

Thunder from the Pulpit

1. Material by and about American religion is voluminous. A good beginning source is the National Council of Churches, 475 Riverside Drive, New York, NY 10115-0050, and the Christian Broadcasting Network, CBN Center, Centerville Turnpike, Virginia Beach, VA, 23463. Chapter 3 in Merle Curti's *The Growth of American Thought* (New York: Harper and Brothers, 1943) is still one of the best summaries. This can also be said about William W. Sweet's *American Culture and Religion: Six Essays* (New York, Cooper Square, 1972). Of great use to me is Russel B. Nye's *The Almost Chosen People: Essays in the History of American Ideas* (East Lansing: Michigan State University Press, 1966). It supplied the framework for my book *Great Awakenings: Popular Religion and Popular Culture* (Binghamton, NY: The Haworth Press, 1995). It contains a long section of "Further Readings."

Living with Machines

1. An excellent summary of the dilemma appears in John F. Kasson's *Civilizing the Machine,* and in many books by Lewis Mumford. See, for example, his *Technics and Civilization* (1934), *The Condition of Man* (1944), *Values for Survival* (1960), *The City in History* (1961), *The Myth of the Machine* (1967), and *The Pentagon of Power* (1970). See also Jacques Ellul, *The Technological Society* (1964).

2. These matters are fully explored in Theodore Roszak's *The Making of a Counter Culture* (New York: Random House, 1986).

3. See Roland Gelatt, *The Fabulous Phonograph, 1877-1977* (1977).

4. Many details are given in Carl Belz's *The Story of Rock* (1972), and H. Kandy Rohde's *The Gold of Rock & Roll, 1955-1967* (1970).

5. Caroline Ware, *The Twentieth Century* (New York: Harper & Row, 1966), p. 75.

Index

Plato, 15, 44
cave image, 244, 252
Pocahontas, 250
Pokemon, 196
Polarization, 176
Poor Richard's Almanac (Franklin), 20
"Pop"
American artists, 205
definitions, 43
as derogation, 214, 215
Popular culture
and American Studies, 35
analysts of, 19
and architecture, 147
and communications, 49
and democracy, 50
description of, 16, 213
and discontent, 177
frenzies, 27, 30
globalism, 43
history of, 17, 214
and power, 43-44
Reader's Guide, 1960, 23
topics, 47-48
Popular Culture Association, 23, 35, 216
Popular Culture in Early Modern Europe (Burke), 15
Popular Press, the, 24, 25
Populus, 44
Posner, Richard, 72
Postman, Neil, 84, 85, 161
Postmodernism, 58
Powers, Richard Gid, 48
Premodern culture, studies of, 20
Presidential election, 2000, 1-2
Presley, Elvis, 167, 185-187
Primitivism, 147
Principles of Scientific Management, The (Taylor), 84
Probe, definition, 13, 185
Professional Correctness (Fish), 72
Progress, and development, 234-235
"Public intellectuals," 72
Public Intellectuals: A Story of Decline (Posner), 72

Public Opinion (Lippman), 91
Publishing, and popular culture, 216

Radio, 167
Ranch Life and the Hunting Trail (Roosevelt), 105
Read, Herbert, 149
Red River Lumber Company, 114, 115, 116, 117
Redfield, Robert, 15
Religion, 153, 233
Religious gap, 68
Republic, The (Plato), 15, 44
Revivalism, 155
Revolution, changing meaning, 24
Richman, Hyman, 141
Ritzer, George, 97
River Rouge, Ford plant, 135, 136
Rockefeller, John D., 127, 191
Rodchenko, Aleksandr, 206
Rodeo, definition, 106
Rogers, Buck, 161-162
Rogers, Roy, 107
Rogers, Will, 138
Rollin, Roger B., 88
Roosevelt, Franklin D., 2, 167, 173, 196
Roosevelt, Theodore, 105
Roszak, Theodore, 179
Rousseau, Jean-Jacques, 103, 154, 176
Rule, James, 73
Rumsfeld, Donald, 182
Russell, Kitty, 109
Ruth, Babe, 193

Sacks, Peter, 162
Sandburg, Carl, 114
Sarnoff, David, 167
Scandals, 6
Scharnhorst, Gary, 192
Schlesinger. Arthur Jr., 88
Scholarship
democratization, 24
limitations, 216